Physics Teaching and Learning: Challenging the Paradigm

A Volume in:
Research in Science Education (RISE)

Series Editor

Dennis W. Sunal
Cynthia Szymanski Sunal
Emmett L. Wright

Research in Science Education (RISE)

Series Editor

Dennis W. Sunal
Cynthia Szymanski Sunal
University of Alabama

Emmett L. Wright
Kansas State University

Physics Teaching and Learning: Challenging the Paradigm

Edited by

Dennis W. Sunal
Jonathan T. Shemwell
James W. Harrell
Cynthia S. Sunal

INFORMATION AGE PUBLISHING, INC.
Charlotte, NC • www.infoagepub.com

Library of Congress Cataloging-In-Publication Data

The CIP data for this book can be found on the Library of Congress website (loc.gov).

Paperback: 978-1-64113-656-3
Hardcover: 978-1-64113-657-0
E-book: 978-1-64113-658-7

Printed in the United States of America

CONTENTS

PREFACE TO THE SERIES

Science education as a professional field has been changing rapidly over the past two decades. Scholars, administrators, practitioners, and students preparing to become teachers of science find it difficult to keep abreast of relevant and applicable knowledge concerning research, leadership, policy, curricula, teaching, and learning that improve science instruction and student science learning. The literature available reports a broad spectrum of diverse science education research, making the search for valid materials on a specific area time-consuming and tedious.

Science education professionals at all levels need to be able to access a comprehensive, timely, and valid source of knowledge about the emerging body of research, theory, policy and practice in their fields. This body of knowledge would inform researchers about emerging trends in research, research procedures, and technological assistance in key areas of science education. It would inform policy makers in need of information about specific areas in which they make key decisions. It would also help practitioners and students become aware of current research knowledge, policy, and best practice in their fields.

For these reasons, the goal of the book series, *Research in Science Education,* is to provide a comprehensive view of current and emerging knowledge, research strategies, and policy in specific professional fields of science education. This series presents currently unavailable, or difficult to gather, materials from a variety of viewpoints and sources in a usable and organized format.

Physics Teaching and Learning: Challenging the Paradigm, pages vii–viii.
Copyright © 2019 by Information Age Publishing
All rights of reproduction in any form reserved.

Each volume in the series presents a juried, scholarly, and accessible review of research, theory, and/or policy in a specific field of science education, K–16. Topics covered in each volume are determined by current issues and trends, as well as generative themes related to up-to-date research findings and accepted theory. Published volumes will include empirical studies, policy analysis, literature reviews and positing of theoretical and conceptual bases.

PREFACE

Dennis W. Sunal, Jonathan Shemwell,
James W. Harrell, and Cynthia S. Sunal

Physics Teaching and Learning: Challenging the Paradigm, RISE Volume 8, focuses on research contributions challenging the basic assumptions, ways of thinking, and practices commonly accepted in physics education. Teaching physics is a multifaceted, research-based, value added strategy designed to improve academic engagement and depth of learning.

Chapter goals are selected to promote dialogue in the research community for sharing research on teaching conducted in grade level 9–16 classroom contexts. This strategy may include impact of challenging the paradigm in:

- Professional development in teaching physics
- Student learning characteristics
- Effective teaching practices
- Interventions designed to enhance physics learning
- Improving and sustaining professional learning communities among teachers of physics.
- Classroom context, emerging curricular models and technological applications promoting physics learning.

Our nation faces a crisis in availability of quality pre-service and in-service physics teachers. Out-of-field teaching, as indicated through a lack of appropriate

Physics Teaching and Learning: Challenging the Paradigm, pages ix–x.
Copyright © 2019 by Information Age Publishing

preparation in physics content knowledge and/or teacher certification, has a strong potential negative impact on science instruction and student performance. Two-thirds or more of secondary physics teacher fall under this category. A majotity of physics teachers do not have a college major or minor in physics. Inadaquate and low quality teacher preparation programs also affect up to one-half of inservice physics teachers. It is not uncommon for teacher preparation to exclude a student internship or even part time clinical work in classroioms before graduation.

Such negative impact was evidenced by the low to mid-level science scores of U.S. students in the Program for International Student Assessment (PISA) results when compared to the rest of the world. Effective physics teaching requires more than physics knowledge; it also requires strong physics pedagogical content knowledge (PCK), Physics PCK includes knowledge of high school students' proir physics ideas, reasoning patterns, research-based instructional approaches, cultural background, and learning behaviors integrated with knowledge of physics. Experts in business and education argue that unless America renews a commitment to excellence and development of scientific talent, the long-term prosperity of our nation may be at risk.

Having been recognized as an effective strategy for inspiring student learning, innovative and non-standard active learning, and engagement in school, creative physics teaching and learning outside the traditional instructional patterns has been credited for improving high school graduation rates, motivating students to attend and finish college, and engage in careers about which they are passionate.

In the chapters that follow, a number of researchers, teaching and curriculum reformers, and reform implementers discuss a range of important issues. The volume should be considered as a first step in thinking through what physics teaching and physics learning might address in teacher preparation programs, in-service professional development programs, and in classrooms throughout the U.S. To facilitate our thinking about research-based physics teaching and learning each chapter in the volume was organized around five common elements:

1. A significant review of research in the issue or problem area.
2. Themes addressed are relevant for the teaching and learning of K–16 science.
3. Discussion of original research by the author(s) addressing the major theme of the chapter.
4. Bridge gaps between theory and practice and/or research and practice.
5. Concerns and needs are addressed of school/community context stakeholders including students, teachers, parents, administrators, and community members.

—*October, 2018*

CHAPTER 1

HIGH SCHOOL PHYSICS TEACHING REFORM

Support for Professional Development in the Literature

Cynthia S. Sunal, Dennis W. Sunal,
Justina Ogodo, and Marilyn Stephens

Teacher professional development is often a neglected component of creating reform in classrooms. We examine the literature relevant to physics focused in-service professional development. Considered are studies aimed at addressing the needs of teachers of physics in developing and implementing inquiry-oriented reform teaching strategies. Also examined is the instrumentation used to investigate efforts for the professional development of physics teachers. Elements have been identified which support physics evaluation of teacher development using quantitative, qualitative and mixed method approaches. These studies indicate instrumentation is available for the investigation of physics teacher professional development. We find further investigation, particularly long-term studies, of teachers from diverse contexts is needed.

Keywords: professional development, instrumentation, classroom reform, physics education, student learning, instructional practice

Physics Teaching and Learning: Challenging the Paradigm, pages 1–20.

Concerns have been expressed over the need for high quality physics teachers and course offerings (Banilower, 2013; Hodapp, Hehn, & Hein, 2009; Paek, Ponte, Sigel, Braun, & Powers, 2005; Sunal, Dantzler, Sunal, Turner, Harrell, Aggarwal, & Simon, 2016; White & Tesfaye, 2014). While concerns have been expressed over recent decades, growth in the number of physics teachers has not kept pace with the record enrollments of students predicted by the National Center for Educational Statistics (Dunlop Velez, 2015; Goldring, Gray, & Bitterman, 2013; Ingersoll, 2003; Meltzer, Plisch, & Vokos, 2012; NCES, 2009; Tesfaye & White, 2012). National surveys conducted over the past decade focused on describing the status of physics teachers and physics students. A slight growth in physics enrollments from the 1.35 million students in 2008–2009 to 1.38 million in the 2012–2013 school years was identified by White and Tesfaye (2014). This finding also was associated with a decrease of students in regular physics classes and an increase of 125,000 in other specialized physics classes, such as honors and Advanced Placement (AP) physics, pre AP, and International Baccalaureate (IB) physics classes. This trend is currently increasing and challenging the paradigm of what it means to become a physics teacher. In a national survey, White and Tyler (2014) found that although the number of students taking physics increased slightly, the number of teachers of physics and, of qualified physics teachers within that number, remained unchanged from previous years. Only about 27,000 teachers taught at least one physics class during the 2012–2013 school year. Studies have observed the popularity of reform and new paradigms in physics teaching and learning such as conceptual physics, use of mathematical modelling in instruction, student inquiry as a learning strategy, advanced placement physics, and of physics as a high school graduation requirement has driven physics enrollment such that, now, the challenge is how to fill the classrooms with sufficient numbers of qualified or specialized physics teachers (Meltzer & Otero, 2015). The evident concerns about the professional development of teachers qualified to teach physics in the secondary school leads to the research questions around which this chapter is focused. What does the research literature tell us about supportive professional development strategies addressing the diverse needs of teachers of physics in developing and implementing reform teaching in classrooms? What instrumentation is being used to investigate efforts for the professional development of teachers of physics?

BACKGROUND

While a need for qualified physics teachers has been identified, the NCES (Dunlop Velez, 2015) reported approximately 65% of earth science and chemistry, and more than 70% of physics and physical science classes are being taught by teachers who do not have a major or minor in this teaching assignment. Hence, much secondary school science, including physics, is taught in the United States, when taught at all, by teachers considered to be out-of-field for teaching science. Out-of-field teaching is defined as instruction by teachers who do not have teacher certification in their main teaching assignment (Gray & Taie, 2015; Seastrom, 2004). Such out-of-field instruction is particularly common in physics where 13.7% of the teachers were reported in 2015 as not certified in teaching science or in any teacher education field (NCES). Lack of certification has a potential negative impact on science instruction and student performance as evidenced by low to mid-level science scores of U.S. students on the Program for International Student Assessment (PISA) when compared to the rest of the world (Breakspear, 2012, 2017).

A report by the Task Force on Teacher Education in Physics (T-TEP) (Tesfaye & White, 2012) states that only 35% of physics teachers in the United States have a degree in physics, which explains why "many current science teachers lack the content knowledge and focused pedagogical preparation needed to teach physics effectively" (Meltzer, Plisch, & Vokos, 2012, p. 1). Other studies found only 10–30% of physics teachers nationally were adequately prepared to teach physics due to lack of a major or of teacher certification (Banilower, 2013; Hoddap, Hehn, & Hein, 2009; Sunal et al., 2016). Although there are many highly qualified physics teachers, the overall situation for U.S. high-school physics students is not good, and

> except for a handful of isolated pockets of excellence, the national landscape of physics teacher preparation shows a system that is largely inefficient, mostly incoherent, and completely unprepared to deal with the current and future needs of the nation's students (Meltzer, Plisch, & Vokos, 2012, p. 8).

Physics education programs, according to Meltzer, Plisch, and Vokos (2012, p. xii) also do "little to develop physics specific pedagogical expertise of teachers." In addition, Meltzer and Otero (2015) indicate being an effective physics teacher requires more than knowledge of physics; it also requires knowledge of high school students' specific physics prior knowledge, reasoning patterns, and learning behaviors. Physics teachers must be able to design and execute activities appropriate for guiding learning and assessing their students. A survey conducted by Banilower (2013) to describe the status of high school physics, reported traditional lectures and quantitative problem solving were most frequently used by physics teachers and physics teachers were more likely to teach multiple subjects than other science teachers. Banilower concluded physics teachers were somewhat less qualified to teach the subject than were teachers of other sciences and were, therefore, in need of professional development. The publication of *America's Lab Report: Investigations in High School Science* (National Academy of Engineering, 2005) reported the need to make physics more of a laboratory-based science, more interesting and more inspiring for students. Meltzer and Otero (2015, p. 8) noted the NRC report drew "attention to the preparation of teachers to facilitate investigation-based laboratory work in the classroom" as well as "increased engagement of students in science laboratory activity."

CREATING REFORM IN HIGH SCHOOL PHYSICS CLASSROOMS

Over the past 30 years national school science education policy and reform efforts through systemic change were developed partly via reports including *Benchmarks for Science Literacy* (American Association for the Advancement of Science, [AAAS], 1993), *National Science Education Standards* (NSES) and National Research Council, 2013), *Science for All Americans* (AAAS, 1990), *Inquiry and the National Science Education Standards: A Guide for Teaching and Learning* (National Research Council, 2000), A *Framework for K–12 Science Education: Practices, Crosscutting Concepts, and Core Ideas* (National Research Council, 2011), and, most recently, through the current science education standards *Next Generation Science Standards* (NGSS) (NRC, 2013). The basis of current reform includes student inquiry as a key pedagogical method. Using inquiry for learning science was advocated in the early 20th century and strengthened in these later documents. John Dewey, in 1909, critiqued classroom science teaching as having too much emphasis on the accumulation of information and not enough on science as a way of thinking as described in current thinking (NRC, 2000). Since the progressive era science education reform has focused on the shift from teacher-centered to learner-centered approaches in the classroom. Dewey (1938) introduced "learning science by doing" a learning strategy that focuses on practical experience and skills to promote students' conceptual learning. Science teachers, more recently, were expected to possess content knowledge integrated with appropriate pedagogical knowledge, creating a new knowledge organization identified as pedagogical content knowledge, considered necessary for effective science disciplinary teaching (Shulman, 1986, 1987). *Inquiry and the National Science Education Standards* (NRC) highlighted the key features of inquiry placing the responsibility of implementing reform practices on the teacher: 1) the learner engages in scientifically oriented questions; 2) the learner gives priority to evidence in responding to questions; 3) the learner formulates explanations from evidence; 4) the learner connects explanations to scientific knowledge; and 5) the learner communicates and justifies explanations to enhance teachers' content knowledge and instructional practices.

EFFECTIVE PHYSICS FOCUSED PROFESSIONAL DEVELOPMENT

Physics teacher knowledge is a key element to be considered in professional development (PD) aiming at supporting the understanding, implementation, and performance of high quality physics teaching. Typical in-service teacher PD doesn't integrate subject matter and pedagogy (Lederman & Gess-Newsome, 1999; Smith, 1999). The integration of physics content with high levels of physics pedagogical content knowledge (PCK) is a critical factor needed to create meaningful physics learning outcomes. In addition to the theoretical knowledge from professional development experiences, there was a need for clinical experience practice for in-service teachers. Clinical practice, it was argued, would connect theory to practice. Darling-Hammond (2010) states "powerful teacher education programs should have a clinical curriculum as well as a didactic curriculum...

no amount of coursework can, by itself, counteract the powerful experiential lessons that shape what teachers actually do" (p. 40, 43). Subject matter structures used typically by in-service teachers have origins that are only content-oriented yet, physics pedagogical content knowledge (PCK) includes knowledge of how particular physics subject matter topics, problems, and issues can be organized, represented, and adapted to the diverse interests and abilities of learners and presented for instruction (Crouch, Watkins, Fagen & Mazur, 2011; Knight, 2004).

Effective physics focused professional development needs to address research-based strategies found in the literature for effective classroom reform (Guskey, 2000; Guskey & Yoon, 2009, & Loucks-Horsley, Love, Stiles, Mundry, & Hewson, 2003). Such reform strategies encompass lesson planning, implementation, and evaluation. Several guidelines based on research studies exist. High quality professional development was described in the *No Child Left Behind* (NCLB) Act (2001) as 1) sustained, intensive, and classroom-focused; 2) directed to improve and increase teachers' academic subject knowledge; and 3) directed to increase teaching skills by substantially advancing teachers' understanding of effective instructional strategies. The NCLB Act utilized Shulman's (1986) concept of pedagogical content knowledge to examine how professional development enhances teachers' knowledge base to produce meaningful learning among students. High quality professional development aligns with the goals and standards set by the state or district and provides "opportunities for collaboration so that teachers can learn from each other" (Cochrane-Smith, 2004, p. 49).

Five key professional development elements were identified by Desimone (2011) to evaluate the effectiveness of physics focused professional development on in-service teacher participants and subsequently on students' learning. The elements included 1) having a content focus, 2) engaging participants in active learning, 3) providing sufficient duration and spacing of training and practice, 4) emphasizing coherence in training and classroom practice, and 5) supporting collaboration of efforts among peer teachers. The presence and extent of these key elements in a professional development experience are related to greater fidelity of reform classroom practice by teachers. The five elements appear to be necessary to understand how professional development works (Garet, Porter, Desimone, Birman, & Yoon, 2001; Wayne, Yoon, Zhu, Cronen, & Garet, 2008). In a survey conducted to determine if features of professional development programs were influencing participants' classroom practices and student outcomes, O'Brien and McIntyre (2011) found all five elements of professional development prescribed by Desimone were needed in the professional development programs experienced by their participants. There was evidence that the professional development program had positive effects on teachers' classroom practice and student learning outcomes (Blank & de las Alas, 2009; Yoon, Duncan, Lee, Scarloss, & Shapley, 2007).

Content Focus

Physics focused professional development provides teachers with disciplinary content knowledge and with knowledge of how secondary students learn that content. In describing professional development that works, Birman, Desimone, Porter, and Garet (2000) concluded "the degree to which professional development focuses on discipline content knowledge is directly related to teachers' reported increase in depth of knowledge and skills" (p. 30). Other empirical research also suggests content focused professional development integrates teachers' knowledge and teaching practice with increased student learning (Desimone, Smith & Ueno, 2006, Ingvarson, Meiers, & Beavis, 2005; Rock, Courtney, & Handwerk, 2009). Using inquiry learning instruction moves the teacher's role from being the center of knowledge in lecture-based instruction to being a guide and facilitator who uses collaborative groups to engage students in explaining, clarifying, and justifying what they have learned while the teacher listens and encourages broad participation, and uses the information to form the core of classroom instruction (National Research Council, 1996). Because of the complex and sophisticated nature of inquiry, there is a need for a strong emphasis on significant professional development and continuous support of teachers (Capps & Crawford, 2009). Curricular changes towards more inquiry and more student centered classrooms are seen as goals for reform in science classrooms. These changes, for example, require teachers to understand students' prior knowledge and level of comprehension of a science concept and then extend that knowledge into instruction providing active learning, such as a discrepant event, that challenges those prior ideas (National Research Council, 2013). To understand how to take students' prior knowledge and transform the content knowledge into pedagogically powerful instruction at different learning levels requires what Shulman termed pedagogical content knowledge or PCK (Park, Jang, Chen, & Jung, 2010; Park & Oliver, 2008).

Despite reform efforts, little has changed concerning how science is taught in a majority of America's classrooms (Capps & Crawford, 2009; Kahle & Woodruff, 2011; Sunal et al. 2016; & Sunal & Wright, 2006). Although inquiry teaching was favored by in-service teachers, it was not extensively implemented because in-service teachers did not understand what inquiry teaching involved, nor how to implement or assess learning under inquiry (Sunal & Wright, 2006; Sunal et al. 2016). Capps and Crawford concluded inadequate teacher preparation was linked to teacher unfamiliarity with inquiry instruction. Typically, because of their unfamiliarity with inquiry, teachers develop hybrid-reform teaching actions structuring inquiry as cookbook style laboratories with hands-on materials (Sunal & Wright, 2006). When examining how inquiry was enacted in secondary science classrooms in six countries, Abd-El-Khalick and Akerson (2004) found diverse conceptions of inquiry among teachers. They noted classroom enactment of inquiry was highly contextualized. Finally, they found most science teachers had never directly experienced authentic scientific inquiry during their science education or within their teacher education programs; therefore, having experienced no appropriate models, they were unable to enact or use inquiry in their classroom. To examine the extent to which teachers' views of inquiry and practice aligned with ideas in reform-based documents, Capps and Crawford (2012) used a national sample of 5th to 9th grade teachers selected across the country. While inquiry was evident in some of the classes observed, an understanding of inquiry was conspicuously absent for all the participants. They noted the lessons observed had "neither explicit nor implicit instruction related to understandings about inquiry" (Capps & Crawford, 2012, p. 510).

For effective reform-oriented teaching to take place in the physics classroom, the teacher must be adequately prepared and equipped in content and integrated with instructional strategies in order to meet the needs of the student. Because many teachers do not experience reform based teaching and the learning that can result, professional development should integrate components that help teachers develop high levels of physics PCK (Etkina, 2010). "Few professional development programs seem to focus on the topic-specific nature of teacher knowledge" (p. 472) noted Zhang, Parker, Koehler, and Eberhardt (2015).

Participant Active Learning

Physics content focused professional development (PD) should involve teachers in active learning, "engaging teachers as learners" (Ingvarson et al., 2005). Teachers need to receive feedback during PD activities and not be passive participants. Ingvarson et al. noted the existence of a positive relationship between active learning during professional development and teachers' knowledge of practice. To understand the effect of different characteristics of an inquiry science professional development program on teachers' knowledge and ability to implement a program in the classroom, Penuel, Fishman, Yamaguchi, and Gallagher (2007) conducted a two-year study. Their results showed participants' knowledge and preparedness toward inquiry were significantly impacted by continuous training because of the core features of the program which focused on "content knowledge, active or inquiry-oriented learning approaches, and a high level of coherence with other reform activities and standards in the teachers' local school context and practice" (p. 924). These factors contributed to enhancing the content knowledge, pedagogical skills, and changes in the participants' instructional practices.

Effective professional development adds to teachers' professional instructional repertoire enhancing PCK as they may develop creativity and intentionality among their diverse students when they differentiate instruction. Increasing a teacher's pedagogical repertoire also increases the use of reform teaching in the classroom. The effect of increasing the physics teachers' repertoire helps the teacher to differentiate how particular physics topics or concepts, problems, and issues are organized, presented, and modified for the different levels of learners in the classroom (Crouch, et al. 2011).

Sufficient Duration and Spacing of Training and Practice

Duration is provision of an adequate period of time for sustained learning. Physics content focused professional development activities should be long enough to provide the teacher with adequate time to utilize the information. A national goal for effective professional development "is to increase responsibility and accountability for professional development programs to better equip teachers to teach a rigorous curriculum to all students and to ensure that students meet high standards" (Mundry & Boethel, 2005, p. 3). This goal can be achieved through ongoing training. The National Research Council (1996) noted "becoming an effective science teacher is a continuous process that stretches from pre-service to the end of professional career" (p. 54) so "teachers will need ongoing opportunities to build their understanding and ability" (p. 56).

Professional development providing 30 to 100 hours of continuous training had a statistically significant and positive effect on student classroom achievement gains compared to those with five to 14 hours according to findings by Yoon, Duncan, Lee, Scarloss and Shapley (2007). In a case study examining the impact of inquiry-based professional development on teachers' core conceptions and teaching practices, Kazempour (2009) followed one teacher's progress in an inquiry based continuous professional development program. The focus of the study was to better understand the experiences, changes in conceptions of teaching, and the factors influencing the classroom practice of participants. The study found effective professional development enhancing and changing a teacher's classroom practice, must 1) be continuous over a period of time, 2) involve teacher active participation in authentic scientific inquiry-based activities and discussions and 3) model effective inquiry-based practices. In another study, Banilower et al. (2013) conducted a national survey of science and mathematics teachers to identify trends in 1) teacher background and experience, 2) curriculum and instruction and 3) availability and use of instructional resources. The survey asked about the total amount of time participants had spent on professional development related to their content area. About 30% of middle and high school science teachers had participated in more than 35 hours of content focused professional development in the previous three years. It was concluded that a "brief exposure of a few hours over several years is not likely to be sufficient to enhance teachers' knowledge and skills in meaningful ways" (Banilower, et al., p. 34).

Coherence in Training and Classroom Practice

Coherence is the extent to which professional development is consistent with teacher learning opportunities, with teachers' knowledge and beliefs, and with teachers' school, district, and state reforms and policies (Desimone, 2011). Coherent physics focused professional development activities are aligned and consistent with the teachers' goals and assessments, and promote discourse among teachers concerning their work. In a case study to identify the degree to which participating in a three-year professional development program affected participants' content knowledge and use of inquiry, Jeanpierre, Oberhauser, and Freeman (2005) found teachers translated their experiences into instructional practices and increased their science content understanding. The researchers noted several key characteristics of the professional development allowed teachers to successfully translate gains in professional and content knowledge to their classrooms. The characteristics identified included increase in science content and process knowledge, opportunities for practice, and the requirement that teachers demonstrate competence in a tangible and assessable way.

School context and the coherence of the PD training can moderate the efficacy of instructional practices. Coherence with school context is important in understanding the nature of, and establishing links between, teacher practice and student achievement (Peak et al., 2005). Because teacher experiences are set in contexts in which these forces influence their professional beliefs and teaching practice, it is necessary to consider contextual variables when evaluating the impact of effective professional development (Kang, Cha, & Ha, 2013). Because teachers often work within a broader contextual framework beyond their classrooms, such as the school, the effectiveness of professional development interventions, including their design and implementation, should be described within the context of functioning school settings (Cole, 2004; Yoon, 2008). In addition to the classroom physics setting and the characteristics of students, contextual factors can include variables representing reasons for lack of successful implementation such as insufficient time allotted for science instruction by school sites and districts, mandates to teach a certain curriculum, lack of resources, and classroom management issues (Buczynski & Hansen, 2010; Lee & Zuze, 2011; Schleicher, 2012). Alignment of prescribed methods and strategies with the structure and organization of the classroom, school, and district facilitates an increased acceptance of ideas from the training (Paek et al., 2005; Martin, Mullis, Foy, & Stanco, 2012; Saunders, 2014; Trends in International Mathematics and Science Study [TIMSS], 2012). Teacher professional learning should be conceptualized to reflect "the complex teaching and learning environments in which teachers live" (Opfer & Pedder, 2011, p. 377). When professional development reform is not aligned with the school context, teachers tend not to adopt and implement improved teaching in their classrooms (Desimone et al., 2002; Saunders, 2014). Knowledge from professional training alone, however, is not enough to make desired changes in the classroom. Fullan (2007) noted "the notion that external ideas alone will result in changes in the classroom and school is deeply flawed as a theory of action" (p. 35).

Collaboration of Efforts among Peer Teachers

Collaboration of efforts of teachers in professional development provides opportunities for teachers to work with their colleagues within the context of the face-to-face professional development sessions. This collaboration fosters professional learning communities (PLCs). The PLCs continue to function when the teachers move back into their individual classrooms. As physics teachers are typically single teachers in each school, the PLCs become virtual where face-to-face contact between teachers occurs on social media, reading each other's websites, and through texting and email. A positive association has been found between collective participation through professional development and classroom teaching practice (Desimone, 2011; Meltzer, Plisch, & Vokos, 2012; Penuel et al., 2007).

CONCLUSION

A literature exists examining research on physics teaching and, more specifically, on professional development of in-service physics teachers. The previous discussion indicates several components are necessary if professional development is to be effective. Effectiveness generally is considered in terms of implementation of reform teaching practices aimed at engaging our students in active inquiry learning. We also find the teacher's content knowledge in physics is important. Teachers need a deep understanding of physics concepts if they are to guide students' inquiry into those concepts. Hence, researchers have investigated fostering concurrently the growth and integration of both teaching practices and teachers' content knowledge. The instrumentation supporting the methods used in the investigation of professional development of physics teachers is varied. We next discuss instrumentation utilized by researchers to collect evidence when investigating physics focused teacher professional development.

Instrumentation Used to Gather Evidence in Physics Teaching Reform Studies

Typically research in reformed science teaching, as in physics teaching, and related professional development occur through the conceptual change and the socialcultural traditions (Anderson, 2007). Conceptual change approaches research in science education from the view that student learning in science comes about from a systematic transition from prior, often inaccurate, conceptual understandings to scientifically accepted understandings. These reforms begin with a constructivist, inquiry-based environment and demonstrate a shift from teacher-centered to student-centered teaching and learning (MacIsaac & Falconer, 2002). Physics teaching professional development can be investigated to examine how well this shift is supported by the PD.

Sociocultural approaches science education research from the view that learning in science is primarily influenced by culture and interaction with others. It holds that science is a community/group endeavor, and experiences providing students with unbiased and culturally conscious opportunities to participate in this community open the possibilities for student future participation in science in both careers and the generalized society. Physics teaching professional development can be investigated to examine how beliefs, context, and culture are supported by the PD.

Sample instruments found implementing these research traditions in the literature have been reported as effective means for investigating, measuring and monitoring professional development focused on reformed physics teaching. The research tradition currently have focused on mixed-menthd designs (Creswell & Plano-Clark, 2011).

Reformed Teaching Observation Protocol

The *Reformed Teaching Observation Protocol* (RTOP) (Sawada, 2002; Sawada, Turley, Falconer, Benford, & Bloom, 2002) has been used to inform teachers of areas where improvement should be made using a quantitative rating and a qualitative description derived from in-class observations conducted by a trained observer. The RTOP was designed by Pilburn, Sawada, Turley, Falconer, Benford ... Judson, (2000) and Sawada et al. as a classroom observational tool to measure the degree of reformed teaching in science classrooms. The developers did NOT presume that reformed instruction is necessarily quality instruction. Rather, we left that as a hypothesis to be examined and *tested* in and across various reformed settings. The instrument is divided into five sections: 1) lesson design and implementation; what the teacher intended to do to support student understanding of the lesson, 2) propositional content knowledge including teacher content knowledge, organization and presentation of material, 3) procedural content knowledge; what the students did, type of instruction used to engage them, 4) classroom culture (communicative interactions); how the teacher facilitated interactions among the students, and 5) classroom culture (student–teacher relationships); the culture of respect and comfort supported by the teacher and learners. Each section contains five items rated from 0 (never occurred) to 4 (very descriptive). The total RTOP rating ranges from 0 to 100, with a higher rating representing teaching that is more reformed.

The RTOP has been validated extensively by multiple research studies and has a reliability rating of $\alpha = 0.954$ (Piburn et al. 2000). In a mixed method study examining the impact of modeling instruction on participating teachers' instructional practices, Barlow, Frick, Barker, and Phelps (2014) purposefully selected nine participants in the project TIME professional development program. Participants' classrooms were observed before and after training with the RTOP instrument. Teachers also were interviewed with an interview protocol instrument. Analysis of quantitative data from the RTOP instrument showed all but two participants demonstrated increased total RTOP scores. This finding was described by the researchers as an increased level of reform demonstrated by the teachers following the TIME professional development experience. The two participants who were low scoring on the RTOP scale cited external challenges preventing the implementation of the gains from the professional development.

In another study, RTOP was used to examine how teachers' beliefs influenced the use of inquiry. Lotter, Rushton, and Singer (2013) observed 36 high school teachers' classrooms using the RTOP after teachers participated in a two-week professional development program. They also used a series of interviews to investigate teachers' beliefs about inquiry instruction. Based on the RTOP rating, Lotter et al. found teachers' enactment of inquiry fell into four levels: integrated, emerging, laboratory-based, and activity-focused. They further found teachers needed a strong conceptual understanding of inquiry-based teaching in order to implement inquiry practices.

In the National Science Foundation funded National Study of Education in Undergraduate Science, RTOP was used to measure extent of reform following extensive professional development with undergraduate faculty (Sunal, Dantzler, Sunal, Turner, Steele,...Zollman, 2014). Thirty five science faculty at 20 higher education institutions were observed in their classrooms over a week period using RTOP by multiple observers. The RTOP ratings were used to determine the depth and sustainability of reform characteristics experienced and continued over time in university science classrooms. Using the RTOP ratings Sunal et al. (2014b) found that "Reform efforts were sustainable with collaborative faculty knowledgeable in pedagogical innovations and administrative support" and "A significantly high level of reform both in quality and quantity was required to develop greater than expected gains in student outcomes" (p. 118).

To determine if PCK was necessary for science reform teaching, Park et al. (2010) used the RTOP in a quantitative study to measure the degree of reform-oriented instruction in the classrooms of seven high school biology teachers. They investigated the correlation between a teacher's PCK level and the level of reform in the classroom using a PCK rubric and found the "level of a teacher's PCK is highly connected with the degree to which his or her instruction is reform-oriented" (p. 252). Studies support the use of the RTOP as a valid instrument in assessing the level of reform-oriented practices in the science classroom.

Science Teaching Efficacy Belief Instrument

The *Science Teaching Efficacy Belief Instrument* (STEBI-A) was created by Riggs and further developed by Riggs and Enochs (1990) to assess in-service science teachers' personal self-efficacy belief. Such a belief has been investigated in several studies of professional development. These studies have considered possible impact of professional development on teacher personal self-efficacy. The STEBI-A has two distinct subscales, personal science teaching efficacy (PSTE) and science teaching outcome expectancy (STOE). The PSTE scale measures teachers' beliefs in their own ability to teach science and has a reliability rating of ($\alpha = 0.92$), while the STOE assesses teachers' beliefs that student learning can be influenced by effective teaching. The STOE has a reliability of $\alpha = 0.77$. The instrument contains 23 items with 10 written in positive language and 13 written in negative language. All items on the instrument are on a five-point Likert scale ranging from 1 (strongly disagree) to 5 (strongly agree). The STEBI-A has been well researched and validated as a reliable tool for examining teachers' self-efficacy beliefs toward the teaching of science.

To examine the impact of standards-based professional development on teacher efficacy and instructional practice, Lakshmana, Heath, Perlmutter, and Elder (2011) conducted a three-year longitudinal study using direct classroom observations to observe changes in participants' practices. Two instruments, STEBI-A and RTOP, were used to assess teacher efficacy and reform-oriented practices of the participants respectively. The researchers found significant growth in teacher self-efficacy and in the extent to which teachers' implemented inquiry-based instruction in the classroom. They also found a positive correlation between changes in the use of inquiry-based instructions and changes in teacher self-efficacy. It was determined that both instruments were useful in examining the professional development impacts under investigation.

In a study assessing elementary teachers' science teaching efficacy, Lumpe, Czerniak, Haney, and Beltyukova (2012) used the STEBI-A to examine how teachers' beliefs about teaching science improved as they participated in a long term professional development program. Participants included 450 elementary school teachers involved in six two-week long summer programs focusing on inquiry-based instruction, science content knowledge, and science process (content focus) activities. Although the teachers showed no gains in outcome expectancy beliefs, there were significantly more positive gains in teacher self-efficacy beliefs as a result of the professional development leading to positive benefits for their students' achievement. The researchers concluded teacher beliefs were positively impacted by the number of hours teachers participated in the research-based professional development program and hours of participation were significantly predictive for students' science achievement.

A person's self-efficacy and beliefs often are based on personal judgments of competence to execute a particular task. This judgment is based on acquired or mastered skills used to affect a desired outcome. Erlich and Russ-Eft (2011) noted it is "one's confidence in engaging in specific activities that contribute toward progress to one's goal" (p. 5). The physics teacher's perceptions of his or her skills and ability can influence the teaching of the subject and students' learning.

Content Representation and the Pedagogical and Professional-Experience Repertoires

The *Content Representation* (CoRe) and the *Pedagogical and Professional-Experience Repertoires* (PaP-eRs) measure teachers' pedagogical content knowledge (PCK) in teaching science (Loughran, Mulhall, & Berry 2012). *CoRe* creates a plan of how a teacher conceptualizes the science content of the participant teachers' pedagogical content knowledge. "CoRe becomes a generalizable form of the particular teachers as it links the how, why, and what of the content to be taught with what they agree to be important in shaping students'' learning and teachers' teaching" (Longhran et al., 2012, p. 17). For each key content lesson idea to be introduced in the lesson, 8 question prompts are asked in an interview. They are, 1) What you intend the students to learn about this idea? 2) Why it is important for students to know this? 3) What else you know about this idea (that you don't intend students to know yet). 4) Difficulties/limitations connected with teaching this idea? 5) What knowledge about students' thinking influences your teaching of this idea? 6) What are other factors that influence your teaching of this idea? 7) What teaching procedures will be used and what are the reasons for using them to engage students with this idea? 8) What are specific ways of assessing students' understanding or confusion around this idea? (Loughran, Mulhall, & Berry, 2012).

The PaP-eRs data provides a narrative that combines the classroom lesson observation data and the CoRe interview data to get a more holistic understanding of the teacher's PCK. The authors noted it is important to consider the context of teaching and learning when determining the validity of the CoRe and PaP-eRs.

Physics professional development has endeavored to support and increase teachers' PCK hence, instrumentation addressing the teaching of a physics concept within a specific context has been of interest to researchers.

To properly document the PCK of four college professors, Garritz, Padilla, Ponce-de-Leon, and Rembado (2007) investigated their ways of thinking about the amount of substance when teaching a chemistry concept. They sought to validate Loughran, Mulhall, & Berry's, (2004) CoRe and PaP-ers methodology in evaluating PCK. After conducting the classroom observations and interviews, they concluded Loughran's method helped to uncover, document and portray the instructors' PCK. They also acknowledged the CoRe's eight central idea questions were useful in identifying what the instructors believed to be the main ideas when teaching particular content.

CONCLUSION

This discussion identifies sample key instrumentation used to investigate efforts for the professional development of teachers of physics. Instrumentation with established reliability and validity has been identified and utilized in reported investigations. The instrumentation used must align with the research questions guiding a specific study. Additional instruments, such individual teacher interviews and student focus group interviews, have been documented that align with quantitative, qualitative and mixed method approaches and the questions leading to those approaches.

DISCUSSION

The literature identifies in-service professional development elements supportive of physics teachers' efforts to implement inquiry-oriented reform teaching strategies in their classrooms. These elements are identified in studies that have used instrumentation consistent with different approaches: quantitative, qualitative or mixed method. Overall, five main elements are identified which were described by Desimone in 2001 and in two following reports by Desimone, Porter, Garet, Yoon and Birman (2002) and Desimone, Smith and Ueno (2006): 1) having a content focus, 2) engaging participants in active learning, 3) providing sufficient duration and spacing of training and practice, 4) emphasizing coherence in training and classroom practice, and 5) supporting collaboration of efforts among peer teachers. Other studies report results consistent with these broad categories. Studies also have begun to investigate more specific components of these elements. Our review indicates long-term professional development is necessary in order to sufficiently support in-service teacher physics pedagogical content knowledge development as it is complex and involves matching physics content knowledge, classroom context, and physics instructional activities.

Professional development should include activities aiming at developing deep understanding of physics content, both because many teachers are ill prepared for understanding such content and also as some who have preparation may not be able to sufficiently keep up with current research developments in the field of physics. Deep understanding is necessary as a teacher plans, implements and differentiates strategies to engage all students in meaningful learning.

More research is needed on how to engage teachers in reform teaching strategies. Such strategies are difficult to implement as there is not a prescriptive set of steps teachers and students follow. The literature suggests there are several strategies involved with teacher decision making in reform classroom contexts. Students used to traditional lecture-oriented teaching with cookbook laboratories may resist inquiry-oriented approaches. So, it is likely there will a learning curve for both teacher and students. Finally, the literature indicates professional development aimed at supporting inquiry-oriented reform in diverse physics classroom contexts needs further investigation of its characteristics and its effects and confirmation of instrumentation useful in such investigations.

REFERENCES

Abd-El-Khalick, F., & Ak erson V. L. (2004). Learning as conceptual change: Factors mediated the development of preservice elementary teachers' views of nature of science. *Science Education, 88,* 785–810.

American Association for the Advancement of Science. (1990). *Science for all Americans.* New York, NY: Oxford University Press.

American Association for the Advancement of Science. (1993). *Benchmarks for science literacy.* New York, NY: Oxford University Press.

Anderson, C. (2007). Perspectives on science learning. In S. Abell & L. Lederman (Eds.), *Handbook of research on science education* (pp. 3–30). New York, NY: Lawrence Erlbaum Associates.

Banilower, E. (2013). *2012 National survey of science and mathematics education.* Chapel Hill, NC: Horizon Research, Inc.

Banilower, E. R., Smith, P. S., Weiss, I. R., Malzahn, K. A., Campbell, K. M., & Weis, A. M. (2013). *Report of the 2012 national survey of science and mathematics education.* Chapel Hill, NC: Horizon Research, Inc.

Barlow, A., Frick, T., Barker, H., & Phelps, A. (2014). Modeling instruction: The impact of professional development on instructional practices. *Science Educator, 23*(1), 4–26.

Birman, B. F, Desimone, L., Porter, A. C., & Garet, M. S. (2000). Designing professional development that works. *Educational Leadership, 57*(8), 28–33.

Blank, R. K., & de las Alas, N. (2009). *Effects of teacher professional development on gains in student achievement: How meta-analysis provides scientific evidence useful to education leaders.* Washington, DC: Council of Chief State School Officers.

Breakspear, S. (2012). *The policy impact of PISA: An exploration of the normative effects of international benchmarking in school system performance.* OECD Education Working Papers, (71), Paris, OECD Publishing. http://dx.doi.org/10.1787/5k9fdfqffr28-en

Breakspear, S. (2017). *Developing agile leaders of learning. Wise Learn/Labs.* Retrieved from file:///D:/25%20NOYCE%20TRACK%202/Article%202017%20Breakspear%20Developing%20Agile%20Leaders.pdf

Buczynski, S., & Hansen, C .B. (2010). Impact of professional development on teacher practice: Uncovering connections. *Teaching and Teacher Education, 26*, 599–607. http://dx.doi.org/10.1016/j.tate.2009.09.006

Capps, D., & Crawford, B. (2013). Inquiry-based instruction and teaching about nature of science: Are they happening? *Journal of Science Teacher Education, 24*(3), 497–526, DOI: 10.1007/s10972-012-9314-z

Cochrane-Smith, M. (May/June 2004). The report of the teaching commission: What's really at risk? *Journal of Teacher Education, 55*(3), 195–200. doi.org/10.1177/0022487104264944

Cole, P. (2004). *Professional development: A great way to avoid change, Paper No. 140.* Melbourne, Australia: Centre for Strategic Education.

Creswell, J. & Plano-Clark, V. (2011). *Designing and conducting mixed methods research.* Thousand Oaks, CA: Sage.

Crouch, C., Watkins, J., Fagen, A., & Mazur, E. (2011). Peer instruction: Engaging students one-on-one, all at once. In E. Redish & P. Cooney (Eds.), *Research-based reform of university physics.* College Park: MD, American Association of Physics Teachers, Reviews in PER Vol. 1, Reviews in PER Vol. 1. Retrieved from http://www.per- central.org/document/ServeFile.cfm?ID=4990

Darling-Hammond, L. (2010). Teacher education and the American future. *Journal of Teacher Education, 61*(1–2), 35–47.

Desimone, L. M. (2011). A primer on effective professional development. *Phi Delta Kappan, 92*(6), 68–71.

Desimone, L., Porter, A. C., Garet, M., Yoon, K. S., & Birman, B. (2002). Effects of professional development on teachers' instruction: Results from a three-year study. *Educational Evaluation and Policy Analysis, 24*(2), 81–112.

Desimone, L. M., Smith, T. M., & Ueno, K. (2006). Are teachers who need sustained content-focused professional development getting it? An administrator's dilemma. *Educational Administration Quarterly, 42*(2), 179–215.

Dewey, J. (1909). *Moral principles in education.* Boston, MA: Houghton Mifflin Co. Retrieved from http://www.gutenberg.org/files/25172/25172-h/25172-h.htm

Dewey, J. (1938). *Experience and education.* The Kappa Delta Pi lecture series. Retrieved from https://books.google.com/books?id=JhjPK4FKpCcC&printsec=frontcover&dq=bibliogroup:%22The+Kappa+Delta+Pi+lecture+series%22&hl=en&sa=X&ved=0ahUKEwjX--iwlKLdAhVCJKwKHQQkAV4Q6AEIJzAA#v=onepage&q&f=false

Dunlop Velez, E. (2015). *The condition of education, 2015.* Washington, DC: National Center for Education Statistics.

Erlich, R., & Russ-Eft, D. (2011). Applying social cognitive theory to academic advising to assess student learning outcomes. *NACADA Journal, 31*(2), 5–15.

Etkina, E. (2010). Pedagogical content knowledge and preparation of high school physics teachers. *Physical Review Special Topics Physics Education Research, 6*, 020110.

Fullan, M. (2007). Change the terms for teacher learning. *Journal of Staff Development, 28*(3), 35–36.

Garet, M. S., Porter, A. C., Desimone, L., Birman, B. F., & Yoon, K. S. (2001). What makes professional development effective? Results from a national sample of teachers. *American Educational Research Journal, 38*(4), 915–945.

Garritz, A., Padilla, K., Ponce-de-Leon, A., & Rembado, F. (2007). *The pedagogical content knowledge of Latin-American chemistry professors on the magnitude "amount of sub-*

stance" and it's unit "mole." Paper presented at NARST 2007 Annual Meeting New Orleans, LA.

Goldring, R., Gray, L., & Bitterman, A. (2013). *Characteristics of public and private elementary and secondary school teachers in the United States: Results from the 2011–12 schools and staffing survey. First Look. NCES 2013–314.* Washington, DC: National Center for Education Statistics.

Gray, L., & Taie, S. (2015). Public school teacher attrition and mobility in the first five years: Results from the first through fifth waves of the 2007–08 beginning teacher longitudinal study. *First Look.* NCES 2015–337. Washington, DC: National Center for Education Statistics.

Guskey, T. (2000). *Evaluating professional development.* Thousand Oaks, CA: Corwin.

Guskey, T. R., & Yoon, K. S. (2009). What works in professional development? *Phi Delta Kappan, 90*(7), 495–500.

Hodapp, T., Hehn, J., & Hein, W. (2009) Preparing high-school physics teachers. *Physics Today, 62*(2), 40–45.

Ingersoll, R. M. (2001). Teacher turnover and teacher shortages: An organizational analysis. *American Educational Research Journal, 38*(3), 499–534.

Ingvarson, L., Meiers, M., & Beavis, A. (2005). Factors affecting the impact of professional development programs on teachers' knowledge, practice, student outcomes & efficacy. *Education Policy Analysis Archives, 13*(10), 1–28.

Jeanpierre, B., Oberhauser, K., & Freeman, C. (2005). Characteristics of professional development that effect change in secondary science teachers' classroom practices. *Journal of Research in Science Teaching, 42* (6), 668–690.

Kahle, J. B., & Woodruff, S. B. (2011). Science teacher education research and policy: Are they connected? In G. DeBoer (Ed.), *Research in science education: Vol. 5, The role of public policy in K–12 science education* (pp. 47–75). Greenwich, CT: Information Age Publishing.

Kang, H., Cha, J., & Ha, B. (2013). What should we consider in teachers' professional development impact studies? Based on the conceptual framework of Desimone. *Creative Education, 4,* 11–18. doi: 10.4236/ce.2013.44A003

Kazempour, M. (2009). Impact of inquiry-based professional development on core conceptions and teaching practices: A case study. *Science Educator, 18,* 56–68. http://www.nsela.org/index.php?option=com_content&view=category&id=51&Itemid=85

Knight, R. D. (2004). *Five easy lessons: Strategies for successful physics teaching.* Boston, MA: Addison Wesley.

Lakshmanan, A., Heath, B. P., Perlmutter, A., & Elder, M. (2011). The impact of science content and professional learning communities on science teaching efficacy and standards-based instruction. *Journal of Research in Science Teaching, 48*(5), 534–551.

Lederman, N., & Gess-Newsome, J. (1999). Reconceptualizing secondary science teacher education. In J. Gess-Newsome & N. G. Lederman (Eds.), *Examining pedagogical content knowledge* (pp. 199–214). Boston, MA: Kluwer.

Lee, V. & Zuze, T. (August 2011). School resources and academic performance in Sub-Saharan Africa. *Comparative Education Review 55*(3), 369–397. https://doi.org/10.1086/660157

Lotter, C., Rushton, G. T., & Singer, J. (2013). Teacher enactment patterns: How can we help move all teachers to reform--based inquiry practice through professional development? *Journal of Science Teacher Education, 24*(8), 1263–1291.

Loucks-Horsley, S., Love, N., Stiles, K., Mundry, S., & Hewson, P. (2003). *Designing professional development for teachers of science and mathematics.* Thousand Oaks, CA: Corwin Press, Inc.

Loughran, J., Berry, A., & Mulhall, P. (2012). *Portraying PCK. Portraying PCK. Understanding and developing science teachers' pedagogical content knowledge* (pp. 15–23). Rotterdam, Netherlands: Sense Publishers.

Loughran, J., Mulhall, P., & Berry, A. (2004). In search of pedagogical knowledge in science: Developing ways of articulating and documenting professional practice. *Journal of Research in Science Teaching. 41*(4), 370–391

Loughran, J., Berry, A., & Mulhall, P. (2012). Portraying PCK. *Understanding and developing science teachers' Pedagogical Content Knowledge* (2nd ed., pp. 15–23). Rotterdam, Netherlands: Sense Publishers.

Lumpe, A., Cherniak, C., Haney, J., & Beltyukova, S. (2012). Beliefs about teaching science: The relationship between elementary teachers' participation in professional development and student achievement. *International Journal of Science Education 34*(2),153–166. DOI: 10.1080/09500693.2010.551222

MacIsaac, D., & Falconer, K. (2002). Reforming physics instruction via RTOP. *The Physics Teacher, 40* (November), 16–21.

Martin, M. O., Mullis, I. V. S., Foy, P., & Stanco, G.M. (2012). *TIMSS 2011 international results in science.* Chestnut Hill, MA: TIMSS & PIRLS International Study Center, Boston College.

Meltzer, D. E., Plisch, M., & Vokos, S. (2012). *The role of physics departments in high school teacher education.* College Park, MD: American Physical Society.

Meltzer, D. E., & Otero, V. K. (2015). A brief history of physics education in the United States. *American Journal of Physics, 83*(5), 447–458.

Mundry, S., & Boethel, M. (2005). *What experience has taught us about professional development.* The Eisenhower Mathematics and Science Consortia and Clearing House Network. http://www.sedl.org/pubs/ms90/experience_pd.pdf

National Academy of Engineering. (2005). *America's lab report: Investigations in high school science.* Washington, DC:National Academies Press.

National Center for Education Statistics. (2009). What are the recent trends in advanced mathematics and science course taking among U.S. high school students? *Fast Facts.* Retrieved from http://nces.ed.gov/fastfacts/display.asp?id=97.

National Research Council. (1996). *National science education standards.* Washington, DC: The National Academies Press. https://doi.org/10.17226/4962

National Research Council. (2000). *Inquiry and the national science standards: A guide for teaching and learning.* Washington,DC: National Academies Press.

National Research Council. (2013). *Next generation science standards* (NGSS). Washington D. C: National Academy Press. Retrieved from http://www.nextgenscience.org/

National Research Council, Board of Science Education (2011). *A framework for K–12 science education: Practices, crosscutting concepts, and core ideas* (Common Core). Washington, D.C.: The National Academies Press.

National Research Council (2005). *How students learn: science in the classroom*, Committee on How People Learn, A Targeted Resource for Teachers M. S. Donovan & J.D. Bransford (Eds.). Washington, DC: National Academy Press.

O'Brien, G., & McIntyre, A. (2011). *Designing effective teacher professional learning for improved student outcomes—Research findings from NSW schools.* Paper presented at the ACEL 2011 Annual Conference—Learning Landscapes: Strategies for Sustaining Change, Adelaide, Australia.

Opfer, V., & Pedder, D. (2011). Conceptualizing teacher professional learning. *Review of Educational Research, 81*, 367–407. https://doi.org/10.3102/0034654311413609

Paek, P. L., Ponte, E., Sigel, I., Braun, H., & Powers, D. (2005). A portrait of advanced placement teachers' practices. *ETS Research Report Series, 2005*(1), i–41.

Park, S., Jang, J., Chen, Y., & Jung, J. (2010). Is pedagogical content knowledge (PCK) necessary for reformed science teaching? Evidence from an empirical study. *Research in Science Education, 41*(2), 245–260.

Park, S., &. Oliver, J. S. (2008). Revisiting the conceptualization of pedagogical content knowledge (PCK): PCK as a conceptual tool to understand teachers as professionals. *Research in Science Education, 38*, 261–284. doi:10.1007/s11165-007-9049-6

Penuel, W. R., Fishman, B. J., Yamaguchi, R., & Gallagher, L. P. (2007). What makes professional development effective? Strategies that foster curriculum implementation. *American Educational Research Journal, 44*(4), 921–958.

Piburn, M., Sawada, D., Turley, J., Falconer, K., Benford, R. ... Judson, E. (2000). *Reformed teaching observation protocol (RTOP): Reference manual* (ACEPT Technical Report No. IN00-3). Tempe, AZ: Arizona Collaborative for Excellence in the Preparation of Teachers. (Eric Document Reproduction Services, ED 447 205.)

Riggs, I. M., & Enochs. L. G. (1990). Toward the development of an elementary teacher's science teaching efficacy belief instrument. *Science Education, 74*(6), 625–637.

Rock, J., Courtney, R., & Handwerk, P. (2009). *Supplementing a traditional math curriculum with an inquiry-based curriculum: A pilot of math out of the box.* Princeton, NJ: Educational Testing Service.

Saunders, R. (2014). Effectiveness of research-based teacher professional development. *Australian Journal of Teacher Education, 39*(4), 166–184.

Sawada, D. P. (2002). Reformed teaching observation protocol (RTOP). *School Science and Mathematics, 102*(6), 245–253.

Sawada, D., Turley, J., Falconer, K., Benford, R., & Bloom, I. (2002). Measuring reform practices in science and mathematics classrooms: the reformed teaching observation protocol. *School Science and Mathematics. 102*(6). 245–252.

Schleicher, A. (Ed.) (2012). *Preparing teachers and developing school leaders for the 21st century: Lessons from around the world.* OECD Publishing. http://dx.doi.org/10.1787/9789264xxxxxx-en

Seastrom, M. G. (2004). *Qualifications of the public-school workforce: Prevalence of out-of-field teaching.* U. S. Department of Education. Washington, DC: National Center for Educational Statistics. Retrieved from http://nces.ed.gov/pubs2002/2002603.pdf

Smith, D. C. (1999). Changing our teaching: The role of pedagogical content knowledge in elementary science. In J. Gess-Newsome & N. G. Lederman (Eds.) *Examining pedagogical content knowledge* (pp. 163–198). Boston: Kluwer.

Shulman, L. (1986). Those who understand: Knowledge growth in teaching. *Educational Researcher, 15*(2), 4–14.

Shulman, L. (1987). Knowledge and teaching: Foundations of the new reform. *Harvard Educational Review, 57*(1), 1–23.

Sunal, D., Dantzler, J., Sunal C., Turner, D. Harrell, J.W., Aggarwal, M., & Simon, M. (2016). The 21st century physics classroom: What students, teachers, and classroom observers report. *School Science and Mathematics, 116*(3), 116–126.

Sunal, D., Dantzler, J., Sunal, C., Turner, D., Steele, E., Mason, C., & Zollman, D. (2014). National study of education in undergraduate science: What was learned? In D. Sunal, C. Sunal, E. Wright, C. Mason, & D. Zollman (Eds.), *Research based undergraduate science teaching,* (pp. 67–120). Charlotte, NC: Information Age.

Sunal, D., Sunal, C., Turner, D., Steele, E., Mason, C., Lardy, C., Zollman, D., Matloob-Haghanikar, M., & Murphy, S. (2014a). National study of education in undergraduate science: Research design. In D. Sunal, C. Sunal, E. Wright, C. Mason, & D. Zollman (Eds.), *Research based undergraduate science teaching* (pp. 35–66). Charlotte, N.C.: Information Age.

Sunal, D., & Wright, E. (2006). Teacher perceptions of science standards in K–12 classrooms: An Alabama case study. In Sunal, D. & Wright, E. (Eds*.), The impact of state and national standards on K–12 science teaching* (pp. 123–152). Greenwich, CT: Information Age Publishing.

Tesfaye, C. L., & White, S. (2012). *High school physics teacher preparation.* College Park, MD: American Institute of Physics. Retrieved from https://www.aip.org/statistics/reports/high-school-physics-teacher-preparation.

Yoon, K. S., Duncan, T., Lee, S. W. Y., Scarloss, B., & Shapley, K. L. (2007). *Reviewing the evidence on how teacher professional development affects student achievement. Issues & answers* (REL 2007-No. 033). Austin: TX: Regional Educational Laboratory Southwest (NJ1).

Wayne, A., Yoon, K. Zhu, P. Cronen, S., & Garet, M. (2008). *Experimenting with teacher professional development: Motives and methods.* https://doi.org/10.3102/0013189X08327154

White, S., & Tesfaye, C. I. (2014). *High school physics courses and enrollments—Results from the 1987–2013 nationwide survey of high school physics teachers. Focus On.* December. College Park, MD: American Institute of Physics Statistical Research Center. Retrieved from http://www.aip.org/statistics/trends/reports/highschool3.professional development.

White, S., & Tyler, J. (2014). *Who teaches high school physics? Results from the 2012–2013 nationwide survey of high school physics teachers. Focus On,* December. College Park, MD: American Institute of Physics Statistical Research Center. Retrieved from http://www.aip.org/statistics/trends/reports/highschool3.professional developmentf.

Yoon, K. S., Duncan, T., Lee, S. W. Y., Scarloss, B., & Shapley, K. L. (2007). *Reviewing the evidence on how teacher professional development affects student achievement. Issues & Answers* (REL 2007-No. 033). Austin, TX: Regional Educational Laboratory Southwest (NJ1).

Yoon, S. (2008). Using memes and memetic processes to explain social and conceptual influences on student understanding about complex socio-scientific issues. *Journal of Research in Science Teaching, 45*(8), 900–921.

Zhang, M., Parker, J., Koehler, M. J., & Eberhardt, J. (2015). Understanding in-service science teachers' needs for professional development. *Journal of Science Teacher Education, 26*(5), 471–496.

CHAPTER 2

EFFECTS OF PROFESSIONAL DEVELOPMENT ON REFORM IN HIGH SCHOOL PHYSICS TEACHING

Dennis W. Sunal, Marsha E. Simon, Cynthia S. Sunal,
Justina Ogodo, James W. Harrell, and Mohan Aggarwal

The study investigated the effects of physics focused in-service professional development on classroom reform and student learning among a statewide, diverse sample of teachers. The study included 55 teachers whose classrooms were visited multiple times over two years. Observer visits found significant differences as the professional development progressed in the way physics teachers structured their classrooms, conducted teaching, and engaged students. Increased student learning outcomes were found to be related to the amount of reform implemented. The findings were supported in each of three methodological strands using a concurrent parallel mixed method research design. The results provided a rationale for continued professional development focused on reform in physics teaching among experienced teachers of physics.

Keywords: professional development, classroom reform, physics education, student learning, instructional practices, mixed method design

Physics Teaching and Learning: Challenging the Paradigm, pages 21–56.
Copyright © 2019 by Information Age Publishing
All rights of reproduction in any form reserved.

There is a great need to improve student achievement in science, technology, engineering, and mathematics (STEM). We currently lack large in-depth empirical studies investigating the extent and impact of physics professional development on reform occurring in physics classrooms (Desimone, Porter, Garet, Yoon, & Birman, 2002; Saunders, 2014). To partially address this need, the Alliance for Physics Excellence (APEX) was formed with the goal of transforming physics education at the statewide level. The APEX program was a professional development effort to help teachers gain a deeper knowledge of physics content and employ research-based physics pedagogical strategies to enable students to reach higher achievement levels. The purpose of this study was to assess the level of reform and impact on students demonstrated in classrooms of teachers of physics participating in the APEX physics-focused professional development. Reforms considered classroom context, what physics teaching occurred, and impact of reform efforts on student learning.

REVIEW OF LITERATURE

The effort to develop high quality physics teachers who have content preparation and certification in the field has been documented in recent years with concomitant documentation of a lack of teachers of physics (Banilower, 2013; Hodapp, Hehn, & Hein, 2009; Paek, Ponte, Sigel, Braun, & Powers, 2005; Sunal, et al., 2016; White & Tesfaye, 2010). An increase in the number of students interested in taking a physics course at the secondary school level has been noted along with the lack of qualified teachers (White & Tyler, 2014). Student interest was fueled by the development of conceptual physics and advanced placement courses as well as by, in some states, high school graduation requirements (Meltzer & Otero, 2015).

Professional development programs offer one means of supporting teachers of physics who are delivering instruction with out-of-field preparation and so are lacking in physics content knowledge and physics pedagogical content knowledge (PCK) (Gray & Taie, 2015; Seastrom, 2014). Out-of-field teaching, as indicated through a lack of appropriate teacher certification, has a potential negative impact on science instruction and student performance. Such negative impact was evidenced by the low to mid-level science scores of U.S. students in the Program for International Student Assessment (PISA) when compared to the rest of the world (Breakspear, 2012, 2017). Effective physics teaching requires more than physics knowledge; it also requires knowledge of high school students' specific physics ideas, reasoning patterns, and learning behaviors (Meltzer & Otero, 2015). A survey conducted by Banilower (2013) to describe the status of high school physics, reported traditional lectures and quantitative problem solving were mostly used by physics teachers who also were more likely to teach multiple subjects than other science teachers. Banilower concluded physics teachers were somewhat less qualified to teach the subject than were teachers of other sciences and were, therefore, in greater need of professional development. The aim, then, of physics professional development is to engage teachers in working to develop and implement inquiry-based reform strategies in their classrooms (Meltzer & Otero; Banilower). Such reform strategies are consistent with the *Next Generation Science Standards* (NGSS, National Research Council, 2013).

Five key professional development elements were identified by Desimone (2011) to evaluate the effectiveness of professional development on participants and subsequently on their students' learning: 1) a content-focus, 2) participant active learning 3) sufficient duration and spacing of training and practice, 4) coherence in training and classroom practice, and 5) collaboration of efforts among peer teachers. The presence and extent of these key elements in a professional development experience are related to greater fidelity of classroom reform practice by teachers (O'Brien & McIntyre, 2011).

A more extensive review of the literature and of the instrumentation used by researchers to investigate physics professional development efforts is found in this book's chapter entitled "High school physics teaching reform: Support for professional development in the literature." We now present the APEX program and research investigating its effects.

THE ALLIANCE FOR PHYSICS EXCELLENCE PROFESSIONAL DEVELOPMENT PROGRAM

The Alliance for Physics Excellence (APEX) professional development program was a statewide project funded by the National Science Foundation (NSF) as part of the Mathematics and Science Partnership Program (Sunal, et al., 2014). The professional development program was designed with the goal of transforming physics education in the state of Alabama by enabling in-service physics teachers to acquire a deeper content knowledge of physics and effective physics pedagogical strategies (PCK). All classrooms of the selected teachers were visited to collect data on teaching and learning. This occurred before professional development began, Year 0 data collection. As a result of this needs assessment, the APEX professional development model was adapted to involve highly interactive experiences each summer and during weekends each year, along with continuous online professional learning community involvement over a three-year period. The APEX Model professional development (PD) content units included 39 topics and subtopics and PCK concept goals (see Table 2.1). The APEX model goals, professional development, and curriculum included identifying specific physics PCK teaching strategies; developing inquiry-oriented learning environments; utilizing physics content based on American Association of Physics Teachers (AAPT) units, lesson materials, and resources and Physics Training Resource Agents (PTRA); and identifying barriers to learning such as student alternative conceptions (Druit, 2014) matched to the variety of physics content areas. For example, student alternative conceptions with force and motion involved training with an Internet based instructional tool, *Diagnoser*, designed for teacher and student use to assess prior knowledge and suggest teaching strategies (Thissen-Roe, Hunt, & Minstrell, 2004).

TABLE 2.1. Summary of APEX Professional Development Physics Content Knowledge and Physics Pedagogical Content Knowledge Modules

Year 1	Year 2	Year 3
Content		
• Kinematics	• Fluid Mechanics	• Waves
• Newton's Laws of Motion	• Temperature & Heat	• Sound
• Work, Energy, Power	• Thermodynamics	• Geometrical Optics
• Impulse & Momentum	• Electrostatics	• Physical Optics
• Circular Motion & Rotation	• Conductors & Capacitors	• Atomic Physics
• Oscillations	• CASTLE – Electricity Curriculum	• Nuclear Physics
	• Electromagnetism	
	• Electric Circuits	
	• Magnetic Fields	
Pedagogical Content Knowledge		
• Prior Knowledge and Alternative Ideas	• Prior Knowledge and Alternative Ideas	• Prior Knowledge
• Feedback & Metacognition	• Feedback & Metacognition	• Formative Assessment
• Collaborative Learning	• Collaborative Learning	• Constructivist Learning
• Constructivist Learning	• Constructivist Learning	• Action Research
• Action Research	• Action Research	• Materials Management
• Classroom Management	• Lab Management	• Effective Teaching Strategies
• Effective Teaching Strategies	• Effective Teaching Strategies	• Learning Environments
• Learning Environments	• Learning Environments	• Science Teacher Leadership through professional
	• Science Teacher Leadership role in local schools	conferences, local and state workshops, and leadership roles

Differing from other professional development programs, APEX provided a three-year continuous physics focused program for teachers representing a large diverse statewide population of schools and students. The over-arching premise of the professional development model was to provide research-based integrated content and pedagogical experiences determined through a needs assessment, that afforded teachers with opportunities focused on reform in their classroom instructional practice. Participants were selected to be geographically distributed from across the state and from high schools in the rural, urban, and suburban areas of the state. The professional development model included monitoring participants' progress through pre-, during, and post-training assessment to identify changes in practice and growth in the teachers and to inform ongoing professional development planning and activities. Teachers participated in reform-based activities during the training and were provided with materials and curricular resources to be used in the classroom, such as *Diagnoser*.

Strongly interrelated themes emerged from the APEX, Year 0, needs assessment regarding the lack of 1) adequate physics and mathematics knowledge among teachers; 2) knowledge of mathematical modeling; 3) understanding of inquiry teaching, learning, and assessment; 4) understanding of the importance and relevance of feedback; and 5) professional confidence as a physics teacher. A physics teachers' ability to develop and implement inquiry-oriented lessons and labs, for example, was found to be directly related to depth of reported understanding of the physics and mathematics in a lesson. Without an adequate level of physics content understanding, teachers were not observed planning, teaching, or assessing physics effectively through inquiry.

THEORETICAL MODEL OF THE RESEARCH STUDY

Physics content and pedagogically focused professional development lead to teachers' enhanced content knowledge (CK) and pedagogical knowledge (PK). As the two components improve and are integrated, they lead to enhanced pedagogical content knowledge (PCK) over time with continuous application in practice (Shulman, 1986, 1987). Studies show a direct correlation between teacher PCK and the use of reform practices in the classroom (Desimone, 2011; Loughran, Mulhall, & Berry, 2004, 2008, 2012; National Research Council, 1996; Van Driel & Berry, 2012; Van Duzor, 2011). Teachers, therefore, need adequate PCK to implement reform-oriented instruction in the classroom (see Figure 2.1). The role of professional development is to increase teachers' professional repertoire of practice by improving their content knowledge and pedagogical knowledge which together develop the knowledge base needed to improve classroom instruction.

The type of teacher knowledge developed by physics teachers is a key element that should be considered in professional development attempting to facilitate a high quality of physics teaching. Teacher knowledge structures typically are largely content-oriented and initially formed in undergraduate courses taught by faculty focused on physics research, not on the learning and teaching of physics pedagogical content knowledge (PCK). High levels of physics PCK are a critical factor in creating meaningful physics learning outcomes.

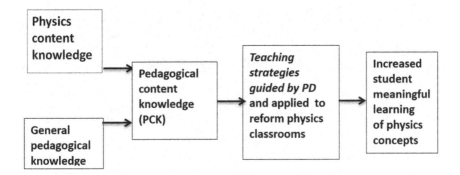

FIGURE 2.1. Knowledge Guiding Teachers in the Physics Classroom

PROBLEM

In studies of in-service teacher professional development, researchers typically note development of the desired integration of subject matter and pedagogy is not accomplished (Abell & Lederman, 2007; Lederman & Gess-Newsome, 1999; National Research Council, 2005; Smith, 1999). Here, PCK includes knowledge of how particular physics subject matter topics, problems, and issues can be organized, represented, and adapted to the diverse interests and abilities of learners and presented for instruction (Crouch & Mazur, 2001; Crouch, Watkins, Fagen

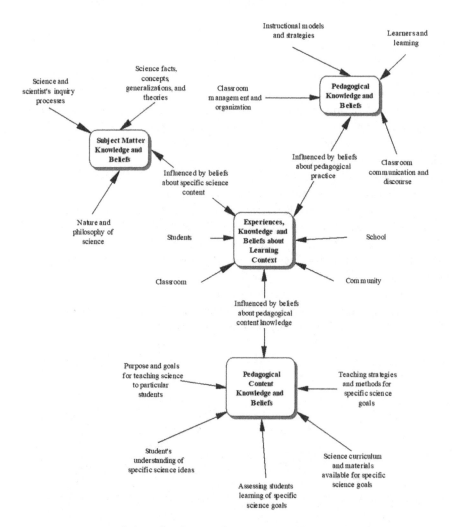

FIGURE 2.2. Knowledge of Pedagogical Strategies Guiding Teachers in the Physics Classroom

& Mazur, 2011; Knight, 2004). Learning in pre-service undergraduate (physics) courses and in-service professional development programs is content oriented, typically reinforcing but not challenging, teachers' fragmented knowledge and relative inability to apply that knowledge within the context of teaching (Lederman & Gess-Newsome).

Our study investigated the development of reform in high school physics classrooms in connection with an extensive professional development program, APEX. The model used in APEX focused on developing teacher PCK in the areas of 1) purpose and goals for teaching physics to particular students; 2) students' prior knowledge of physics concepts, and relationships; 3) teaching strategies and methods for specific physics goals; 4) science curricula content, sequencing, and materials for specific physics goals; and 5) formative and summative assessment of students learning of specific physics goals (see Figure 2.2). We report, here, prior year, first year and second year results.

RESEARCH QUESTION

This study investigated the impact of physics focused professional development on reform in classroom physics teaching occurring in a statewide sample of schools. We examined the overarching research question, "What was the effect of physics focused professional development on the implementation of reform in high school physics classrooms?" To investigate this research question, we considered four sub-questions related to what physics teaching was occurring, the classroom context, and what impact reform was having on teachers and students.

Sub-questions:

1. How have teaching practices changed?
2. What teacher characteristics were related to the implementation of reform practice?
3. What effect did reformed classroom practices have on student learning?
4. How has the classroom learning environment changed?

RESEARCH DESIGN

The APEX professional development program was a partnership between Alabama A&M University, The University of Alabama, and several other regional, state, and national schools, institutions, and agencies. The study was descriptive and predictive with a representative sample of schools and teachers from a diverse, geographically large population of teachers across the state of Alabama. An adaptation of a mixed method design (Creswell & Plano-Clark, 2011) was utilized. The study incorporated a concurrent parallel, exploratory mixed method design with quantitative and qualitative data developed from the viewpoints of teachers, students, and observers. All three strands (teachers, students, and observers) were integrated in a cross strand analysis for all sample high school phys-

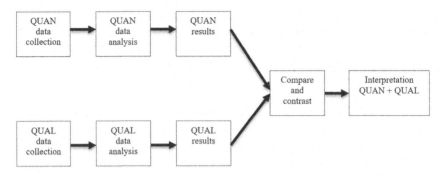

FIGURE 2.3. Triangulation Design: Convergence Model—Concurrent Parallel Mixed Method Research Design

ics teachers, students, and their classrooms. Figure 2.3 illustrates how these data were integrated to produce the results.

PROCEDURE

The study was a part of the Mathematics and Science Partnership Program funded by the National Science Foundation, the Alliance for Physics Excellence (APEX). Critical variables were physics teacher professional development, classroom and teaching reform, and impact of reform variables on physics students. This naturalistic investigation was completed over the 2013–18 school years, looking at physics classrooms prior to and during professional development. The sample physics teachers entered a multi-year intervention program with intensive summer, school year weekends, and continuous online professional learning community involvement.

The study sample included 55 teachers in 55 schools from the Alabama state population of 367 physics teachers in rural, suburban, and urban high schools. Selection criteria included 1) presently teaching one or more physics course(s), 2) balanced representation from each state Intermediate School District, 3) commitment from the school principal of continued assignment to the physics courses, and 4) previous participation in a state physics teacher training program of one to two weeks, Alabama Science in Motion. The majority of schools had only one physics teacher. The high schools ranged from 325 to 2,000+ averaging 1050 students with a high proportion of STEM underrepresented students. The graduation rate of the sample schools was 75.6% with a free or reduced lunch rate of 58.3%. All sample teachers had obtained teacher certification with undergraduate majors in biology/general science (n=47), physics (n=4), or in other STEM fields (n=5). Sample teachers had taught physics an average of 6 years and science for 11 years, and were 54% female (see Table 2.2).

Baseline data, needs assessment data, were collected in the sample physics classrooms in Year 0, before the physics focused professional development in-

TABLE 2.2. Teacher Demographics

Sample physics teachers, in a population of 374, reported:

1. having biology or biology general undergraduate science majors 45 (84%) in college. Four (7%) teachers were physics majors and six (11%) were other majors; chemistry, math, engineering, or earth science.

2. most had teacher certification in general science secondary education with a biology major

3. a wide range of number of years teaching from 1 to 37 years, with an average of 11.5 years in the science classroom and 6.5 years in a physics classroom.

4. most, 32 (58%), were female.

5. many, 70%, had taken at least one university course in methods of science teaching. None had taken a physics methods course. AP teachers had at least one AP physics in-service methods course.

6. types of high school physics classes taught were primarily general physics classes (51%). Other courses taught were AP physics (33%), and honors physics (16%).

7. average number of physics courses taught per day was 1.4 with most teachers, 52.5%, teaching only 1 course.

tervention activities were implemented. During the second year of the three-year intervention, Year 2, a second classroom visit was completed. To increase validity, each of the 55 sample classrooms were visited by trained observers during two consecutive days on two separate occasions, totaling four visits. The classroom visits by the study observers took a total of 134 days to travel to schools located between 23 and 422 miles round trip. During each visit, observers conducted extensive teacher interviews, observed physics lessons and labs, and collected lesson artifacts. Before the visit, teachers completed three online surveys. Physics students taught by each teacher were observed in their classrooms and lab rooms, completed individual surveys, and were interviewed in focus groups.

To address the research question in this study, multiple instruments were used to collect data (see Table 2.3). Observational instruments used during classroom visits included the *Reformed Teaching Observation Protocol* (RTOP) which generates both quantitative and qualitative data used to measure and monitor reformed teaching practices. RTOP was used to evaluate the lessons being taught. The *Student Learning Engagement Rate Protocol* (SLE) was also used by the observer to evaluate student engagement level during the lesson. Quantitative engagement data were collected at regular intervals during the lesson at the same time the RTOP data were collected.

A modified version of the *Science Teaching Efficacy Belief Instrument* survey (STEBI-A) (Riggs & Enochs, 1990) also was used for this study. The word science in the STEBI instrument was replaced with physics to produce the more physics focused *Physics Teaching Efficacy Belief Instrument* (PTEBI) (see Table 2.3). The word science also was replaced with physics in the PSTE and STOE subscales resulting in the use of PPTE and PTOE respectively.

TABLE 2.3. Instruments Used in Data Collection

Teacher Questionnaire Instruments

1. Physics Teaching Efficacy and Beliefs (PTEBI) adapted from Science Teaching Efficacy and Beliefs (STEBI) (Riggs & Enochs, 1990)

2. Teacher Action Research Report (TARR) (Sunal, 2013)

Teacher Observational and Interview Instruments

1. Reformed Teaching Observation Protocol (RTOP) (Sawada, & Pilburn, 2000 and Sawada & Pilburn, 2002)

2. Content Representation (CoRe) (Loughran, Mulhall, & Berry, 2004)

3. Pedagogical and Professional experience Repertoires (PaPers) (Loughran, Mulhall, & Berry, 2004)

1. Teacher Interview (TI) (Sunal, 2013)

Student Questionnaire Instrument

4. Force Concept Inventory (FCI) (Hestines, Wells, Swackhamer, 1992).

Student Observation and Interview Instruments

1. Reformed Teaching Observation Protocol (RTOP) (Sawada, & Pilburn, 2000 and Sawada et. al., 2002), rating of teacher-student interactions and observational narrative section

2. Student Learning Engagement (SLE) (Sunal, 2013)

3. Student Focus Group (SFG) interview focused on the current physics lessons and those previously taught (Sunal, 2013)

Qualitative data were collected during the classroom visits using three instruments. The teacher interview and student focus group interview took place during each two-day visit based on a time and place arranged before the visits. The *Teacher Interview* (TI) protocol focused on teaching physics and the current lessons planned. Part of these questions focused on the teacher PCK and used questions from the *Content Representation (CoRe)* instrument (Loughran, Berry, & Mulhall, 2012.). *CoRe* creates a plan of how a teacher conceptualizes the science content based on the participant teachers' physics pedagogical content knowledge. A final rating of the teacher's physics pedagogical content knowledge included the *Content Representation* (CoRe) section of the teacher interviews and extensive observer lesson observation notes and artifacts, which were consolidated to create the *Pedagogical and Professional-Experience Repertoires* (PaP-ers) PCK rating. The *Student Focus Group* (SFG) interview protocol focused on the current physics lessons and those previously taught. Observers were responsible for transcribing the teacher and student focus group interviews. The observers also completed a *Content Analysis of Artifacts* record sheet using the collected classroom lesson artifacts.

Once the two-day visit was over, observers prepared an *Exit Report* as well as a rating of teacher pedagogical content knowledge using the CoRe and PaP-ers

TABLE 2.4. Teachers' Action Research Report Outline during Force and Motion Unit

Report the results of your action research activity in a written narrative report. The report subtitles include:

1. Description of the school and classroom context of the Force and Motion unit
2. Lesson plans or lesson outlines of unit
3. Brief daily diary of important events that occurred each day
4. Students' pre- and post-test results on the Force Concept Inventory (FCI) test
5. Interview results from a small group of your students
6. Narrative summary of the action research activity. What did you learn?

data based on the teacher interview and lessons observed. Observers conducted member checking with teachers on the CoRe to ensure interview data were correctly represented. A final checklist summary form listing all instruments and data collected from the visit was completed as well.

The data collection instruments used during the visit were selected and piloted based on reported face, construct, and/or predictive validity and checked for observer/rater reliability. Final inter-rater correlation coefficients using classroom observations ranged from 0.78 to 0.94.

As part of the APEX professional development, classroom practice application activities by teachers included conducting and reporting two action research studies in their classrooms during program Years 1 and 2. The focus of each teacher's action research was on the same unit each year, force and motion, and included using the same student achievement pre- and post-test. Although each teacher planned and implemented his or her own unit, teachers were asked to report similar types of information on the unit so they could be compared (see Table 2.4). The Teacher Action Research Reports (TARR) were an additional source of quantitative and qualitative data.

RESULTS

Research Question: *What was the effect of physics focused professional development on the implementation of reform in high school physics classrooms?*

Sub-Question 1: How have teaching practices changed?

Quantitative Data Results from RTOP and SLE

Changes in Reform Classroom Practice

During classroom visit one in baseline Year 0, and visit two in program Year 2 following 1 ½–2 years of PD, the *Refor med Teaching Observation Protocol* (RTOP) (Sawada & Pilburn, 2000) was administered. Descriptive statistics for

TABLE 2.5. Summaries of RTOP Total Rating

RTOP Total Rating Results

RTOP Observation	Mean	SD	Range	Min/Max Score
Year 0 – Pre PD Intervention	50.36	19.59	11.5 - 97	0-100
Year 2 – During PD Intervention	64.67	18.68	37-91	0-100

Teachers' RTOP Total Rating Results Based on Level of Performance

RTOP Level	Traditional	Beginning	Moderate	High
Year 0 - Pre PD Intervention	7	22	14	12
Year 2 – During PD Intervention	1	14	12	28

the RTOP measure are reported in Table 2.5. Classroom visit observations of the teachers were planned as a pre- and post-test with an intervention in between. The RTOP Year 2 mean rating was 64.67 having a range of 36.5 to 91, with 100 being a maximum rating. This rating significantly increased from a baseline Year 0 mean rating of 50.36 with a range of 11.5 to 97. Multivariate analysis revealed significant differences in the overall RTOP ratings. Teachers were rated significantly higher on the RTOP after receiving professional development, $F(1, 108) = 15.35$, $\alpha < .01$. Effect size for this difference was 0.784. The way sample physics teachers, about half way through their 3-year professional development program, structured their physics classroom experiences and conducted teaching was significantly different. MacIsaac and Falconer (2002) report an RTOP rating of 50 represented the presence of some reform classroom characteristics beyond lecture and discussion. An RTOP rating of 65 demonstrated a moderate level of classroom reform with several elements of inquiry clearly present. In this study, for example, reform consisted of more student oriented lessons, labs integrated with lecture, the use of student learning groups, and student argumentation.

Changes in Levels of Reform

The teachers were divided into four performance levels outlined by Budd, van der Hoeven Kraft, McConnell and Vislova (2013) to determine differences along the overall RTOP ratings. The four levels were Traditional, 0–30, Beginning, 31–50, Moderate, 51–70, and High 70–100. More than one-half (n=29) of the sample teachers' classrooms, prior to the PD intervention, Year 0, were not rated as reformed (see Table 2.5). The classrooms were observed to be teacher-centered, using a separate, not an integrated, lecture and lab and lacking a focus on student guided activities or use of mathematical modeling in investigations. Between program Year 0 and Year 2 classroom observations, the number of sample teachers rated as Traditional dropped from seven to one. Most of the rest of the sample teachers were rated as at least partially reformed with the number rated as High

TABLE 2.6, Summary of RTOP Sub-Category Rating Means that Describe the Reformed Physics Classroom

RTOP Observation (SD)	Lesson Design and Implemen- tation	Propositional Knowledge	Procedural Knowledge	Communica- tive Interac- tions	Student/ Teacher Rela- tionships
Year 0					
Pre PD Intervention	8.90 (4.48)	12.17 (3.85)	9.16 (4.97)	9.27 (3.71)	10.84 (4.15)
Year 2 During PD Intervention	12.45* (4.00) +40%	14.16* (3.59) +16%	11.86* (4.15) +29%	12.66* (3.85) +37%	13.64* (4.14) +26%

*Significantly different from pre-Intervention professional development rating

increasing from 12 to 28. Using RTOP ratings, more than one-half of the sample teachers were observed to have reformed their classroom using High rated lessons with a focus on student inquiry.

Changes in Categories of Reform

Reviewing the five RTOP reform sub-categories provided a more detailed measure of specific areas of teacher performance. The five major sub-categories each contain five observed items with a total possible score in each sub-category of 20. Multivariate analysis revealed significant differences in the RTOP subscale ratings (see Table 2.6). Teachers rated significantly higher after receiving professional development in every sub-category, Lesson design and implementation, $F(1, 108) = 19.25$, $\alpha < .01$; Propositional knowledge, $F(1, 108) = 7.34$, $\alpha < .01$; Procedural knowledge, $F(1, 108) = 9.57$, $\alpha < .01$; Communicative interactions, $F(1, 108) = 22.08$, $\alpha < .01$ and Students/Teacher interactions, $F(1, 108) = 12.54$, $\alpha < .01$. There was a significant increase in the level of reform in all five areas of teacher performance among the sample teachers as measured by the RTOP sub-categories. At Year 0 sample teachers were rated two or less on average for each observed item in 4 of the 5 sub-categories. In Year 2 the teachers were rated three or greater on average for each item in all categories. The gains ranged from 16%

TABLE 2.7. Year 0 and Year 2, Levels of Teacher Physics Pedagogical Content Knowledge

Physics Teacher PCK Level		
Yr0	Yr2	PCK Level
08%	38%	Advanced
25%	35%	Proficient
67%	25%	Novice or Emergent

to 40% in the sub-categories. The largest gains were found in Lesson Design and Implementation and in the two Classroom Culture sub-categories.

Using a method reported by Turner and Sunal (2014) with data from teacher interviews and classroom observations, teacher physics PCK ratings before APEX training found 67% of the teachers had a novice or emergent PCK level and 33% were at a proficient or advanced physics PCK level (Table 2.7). Following two years of APEX experience, 25% of teachers were rated at a novice of emergent level of physics PCK and 73% were at a proficient or advanced physics PCK level.

Changes in Student Learning Engagement

Reform also was found in a separate observation made of level of student learning engagement, SLE, a time-on-task instrument. Prior to participation in the in the APEX PD, teachers engaged students in learning the objectives of the lesson about 59.9% (SD 26.16) of the class period, 33 minutes in a 55-minute period. Midway through the PD intervention, in Year 2, teachers were engaging their students in the lesson objectives an average of 76.2% (SD 76.24) of the time, 43 minutes, a 10-minute per day increase. The result was an increase of one additional physics class period experienced by the students per week. A one-way ANOVA comparison of means revealed a significant difference in the SLE ratings from Year 0 to Year 2; $F(1, 92) = 21.18$ at $p<0.01$.

The range of findings among the teachers indicated a need for PD at varying levels. A rating of less than 50 on RTOP and a student active engagement rating of less than 60% of the class time on SLE represents only initial development of physics pedagogical content knowledge (PCK) needed for effective physics teaching on the part of teachers. In other studies these levels did not indicate a reform teacher with more highly developed PCK (MacIsaac & Falconer, 2002).

Qualitative Data Interview Results

Qualitative data were collected and analyzed data from the Year 0 and Year 2 teacher interviews, interviews with small focus groups of each teacher's students, questions from the Content Representation teacher interviews (CoRe), and observers' exit reports. Teacher and student interviews occurred during observers' classroom visits, were recorded, and later transcribed. The observers' finished the Exit Report using these accumulated notes. The summary of data generated the following combined themes from each of these sources.

Year 0, Prior to Intervention Professional Development

Teacher interviews focused on course goals, physics content, teaching methods, and plans for the upcoming lesson to be observed. The most common theme found with pre-intervention interviews in Year 0 was teachers' struggle with their confidence in their understanding of physics content and of ways to accomplish teaching goals. Most teachers felt they were not meeting their goals. They defined their lesson goals as hands-on teaching, by which they meant inquiry teaching. Inquiry was defined differently than is found in the science literature. When asked how to best teach physics, the teachers commonly referred to hands-on learning approaches as "activities," "labs," "problem-solving," "inquiry," "experience," and "discovery" indicating no common name or meaning or sequence. Means of facilitating effective physics learning focused on lower level thinking skills. Skills beyond comprehension and application were rarely mentioned. Observations predominantly found teacher-oriented instruction with lectures and verification labs. Themes developed from the pre intervention Year 0 teacher interviews are summarized in Table 2.8.

When asked about objectives for students in their courses, teachers reported they wanted physics to prepare students for various aspects of post-secondary life ranging from college preparation to real-world application. For example, one teacher intended to prepare students for "calculus based physics in college and to provide a solid foundation." Another asserted, "My goal is that they can apply physics to their everyday life and understand the concept." Most of the teachers in Year 0 wanted students to understand how to practically apply concepts from physics. Some teachers focused on enabling their students to learn enough to meet curriculum and state standards, but for the most part, teachers were interested in students learning physics to achieve success in the future.

Teachers utilized various teaching styles to engage students in physics lessons. Several said they found physics called for instructional methods requiring students to practically apply physics concepts they learned in the course. Teachers, therefore,

TABLE 2.8.   Summary of Year 0 Teacher Interview Themes on Teaching Physics

1.	Deficit in understanding of aspects of physics and mathematics content.
2.	Lack of understanding of inquiry teaching and learning
3.	Understanding of content related to teachers' understanding of inquiry teaching in physics
4.	Teachers strongly cared about student learning of physics
5.	Difficulty in obtaining professional development in physics
6.	Difficulty in implementing differentiated instruction in a physics classroom
7.	Isolation from other physics teachers
8.	Needing awareness of student prior knowledge
9.	Difficulty in developing meaningful student engagement
10.	Difficulty in assessing inquiry learning
11.	Low teacher outcome efficacy
12.	Mistaken understanding of teaching interpreted through biological classroom strategies
13.	Use of outside material and reference resources problematic
14.	Physics seen as a support course, not as amajor
15.	Critical context barriers exist to planning and teaching physics in their schools

designed labs for students to solve problems. For instance, one participant said, "I definitely think labs are important…giving them the opportunity to do things." Many teachers thought student hands-on learning was an effective way for them to comprehend concepts in physics with one teacher mentioning, "Kids have to do hands-on, they have to work through the problems." Many of the teachers asserted the belief that hands-on was the most effective way for students to verify and gain an understanding of the physics concepts. Yet, while they asserted the importance of hands-on learning, teachers reported they first lectured to introduce or explain physics course concepts before lab activities were introduced.

The Year 0 teachers also shared challenges and barriers to physics instruction they experienced in their classrooms. Mathematics and physics are closely related subjects with various concepts in physics necessarily including mathematics. Teachers said some students enrolled in their courses did not have a strong mathematics background so they struggled with having to teach these students mathematics in order to help them to understand physics. A first-year physics teacher mentioned "I guess it is important to know the background of your kids, like mine. Half of mine don't have the math to take the class." Because the students did not possess the required mathematics competency for the course, this teacher said the course had to be "watered down and basic" in order to reach all of the students.

Interviewers inquired about the way in which teachers assessed their students' understanding of concepts during physics class time (formative assessment). Most teachers indicated they assess by spotting the look on their students' faces as they walk around the classroom and ask questions. After being asked about how to determine if a student in their class is confused, one teacher answered, "The looks on their faces [laughs]." In addition to physical indicators of a student's confusion, teachers assessed students' understanding of course concepts through quizzes, tests, and asking questions. One instructor practiced "asking questions…lots of questions!"

Data collected from this sample of teachers illustrated how some held themselves to a standard of teaching aimed at preparing students to comprehend how to practically apply physics in the real-world. Most teachers described important challenges in teaching physics. In addition to students not having the mathematical aptitude for a physics course, teachers also expressed a need for more time in their courses to work with students or to set up labs. One teacher said she did "not have adequate time to set up for their labs" between class periods. As a result, some teachers attempted to implement interactive instructional methods in their courses. Through building a rapport with their students, teachers asked students questions to help them understand physics and connect how physics applied to the larger world outside of the classroom.

Year 2, During Intervention Professional Development

Teacher interviews results during Year 2 professional development were found to be different from those identified in Year 0. Study classroom observers reported teachers were genuinely concerned about their students' learning and understanding of physics. Every teacher observed was described as being interested in helping students gain knowledge about physics concepts. In small focus group interviews, most students rated their teacher's pedagogical approach highly. Some student groups, however, noted they had difficulty understanding their teacher. While mathematics remained an area of concern for teachers and students in Year 2, there were fewer teachers highlighting the challenges of connecting mathematics to physics content.

Lesson Design. In discussing the design of an engaging lesson, all teachers were able to describe elements benefitting student learning, although to different degrees. About a third of teachers were able to articulate reasons why a lesson using a 5E style learning cycle would benefit student learning and engagement. When describing a lesson on the physics concept of work, for example, one teacher engaged students in the following manner.

> So, I included a trick question on a common misconception about what is not work when there is no distance and I said even if you plug the number in there, you still would get zero which shows you there's no work. You can do it either thinking situationally or you can use your math skills. So, it's really just about knowing that they think differently. These boys here love numbers; girls want to see a picture of it, so I did it kind of both ways.

Another third of teachers described how they purposefully left out or added elements of instruction in order to keep students' engaged with the concept. When describing a lesson on the momentum-impulse theorem, for example, a teacher said,

> To me, I care more about them getting it conceptually than the mathematics, for this level. Now, obviously in college, they are going to need the math background so, I don't cut out the math completely, but I care a lot about them conceptually getting it. Because, if they can do the math but don't get why they're doing it, then they're really not doing the math.

Similarly, to engage students on the concept of energy, one teacher said,

> I incorporated various short clips of roller coaster rides that illustrate well the transfer of energy from gravitational potential energy to kinetic energy. Knowing that students have experienced this, and that roller coasters are fun, is a good starting point for this idea.

Developing relationships between variables in a lab setting was a regular in-class activity was discussed by half of the teachers. The 5E learning cycle was seen as a good fit for incorporating instruction around the development of understanding of the relationship among variables in a problem setting.

Among both teachers and students in about one third of the classes, the use of real-world examples during instruction was considered an important element.

Such examples fit well into an inquiry 5E learning cycle but also were sometimes found in traditional lessons including those with a strong use of lecture. One teacher said,

> I try to make connections to things they already know, yes, or try to give them interesting facts maybe, that would make them more interested or ready to learn. I'll show them a picture, something they know, to get them thinking.

Another teacher said, "When I introduce a new topic—I try to relate every topic to something that they're familiar with—like today it was vehicles and cars—and go from there."

Some teachers continued to lecture before doing labs, while most others did some form of inquiry demonstration or lab before doing lecture. One teacher mentioned student-centered learning, but then stated the "guide on the side" approach was not effective because students are not capable of leading themselves to deep understanding of the material. This teacher also expressed a fear that full implementation of the APEX model could jeopardize Advanced Placement (AP) scores. These perspectives and concerns contrasted with what most other teachers said during their interviews.

Formative Assessment: Students' Prior Knowledge and Alternative Conceptions. Many teachers mentioned shaping instruction to address students' prior knowledge or alternative conceptions as described in studies reviewed by Druit (2014). However, only some teachers articulated specific misconceptions that exist for the concept they were teaching. An example of a teacher statement is

> ...but understanding that they have preconceived notions about how energy can move around...they don't realize that they see things in action a lot, so two of the examples I used today were very familiar to them but they don't understand them very well: their car engine and their refrigerator. Those are good examples because they are familiar with those things but they never think about the fact that we take advantage of the principles of heat transfer to make those things work.

Another example statement is

> They hear about light bulbs, and power is measured in watts and so, they don't really think about other connections there. And, they hear the term "doing work" and so, they need to know that there's a physics terminology. And, we use the metric system, you know so they're thinking about watts, they've heard of that but they've never heard of joules. But what they have heard of that similar to joules is horsepower and so, I still need to connect that in because they've heard of horse power, engine horsepower, when they buy a car or something like that. So, I guess it's more learning physics but also the real life applications of what they can connect within their background, horsepower and watts.

Finally, when asked about how teachers' would gauge students' understanding while in class, many teachers described their use of formative assessment, however, some teachers still indicated they would rely solely on students' expressions, questions, or statements in response to their questions. Nearly all teachers used multiple forms of assessment, not leaning on chapter or unit tests as the sole measure of students' understanding of the material taught.

Students' Beliefs about Contributions to Their Learning. Students discussed a variety of classroom elements benefiting their learning. The two most commonly discussed themes were mentioned by half of all of the focus groups. The first of these themes was having a good foundation in mathematics contributes to being able to problem solve in physics. For example, one student mentioned

> I think that it made it easier once we had to start doing the math in there. Because I know in 8th grade we started balancing equations and it was easy for me able to do the math in my head. It was easier for me, and right now that I'm in physics, the math we're doing, I already know how to do it. So, it makes it easier because it works together. We're working at the same pace in math and in physics.

Related to the students' conception of having a strong foundation in mathematics as improving their problem-solving ability, some students suggested having a good foundation in mathematics also improved their attitude towards physics. During one focus group interview, for example, after several students suggested concepts from their mathematics classes were made relevant in their science classes, one student said, "I feel like physics is really easy because math is my best subject and we use a bunch of math in physics, because that's all physics is." Following this comment, another student said,

> I was more open-minded to it, to science. I know I never cared for life science, but once I knew some math was involved in physics and chemistry I was like okay, maybe I'll like this then. And, it is kind of neat to see how the math benefits me in the science classes.

The second main theme regarding contributions to students' learning was inquiry labs and demonstrations helped students pay attention to, and engage with, the physics being taught. This strategy used communication of findings with the whole class in a problem setting through white-boarding. Students were asked to demonstrate their understanding of the relation between variables by reporting their data and graphs and "argue" their case to the class. Public presentations and argumentation was seen as a social event and encouraged and challenged students to work in small groups on gathering data, analysis using the 4-step analysis model and mathematical modeling, discussing the meaning in their group, and reporting the results and relations among variables to the class. Students create meaning and begin to trust inquiry labs as a legitimate way of learning physics content. The theme appears in the following student comments. "When we do labs, it's kind of spontaneous and people weren't expecting it. It gets people to pay attention more." "And, the hands-on stuff is great because if you just sit in front of a book and you're doing all of these word problems, they kind of get boring. So, when you actually put it to test, it's great." "And, not everyone learns that way. Some people need the hands-on."

In a third of the focus groups, students also mentioned they most enjoyed learning how physics applied to their everyday life, a third theme. A fourth theme was evident also in a third of the focus group comments as students reported they enjoyed physics primarily because of their instructor, whether it be a result of their teacher's method of teaching, ability to engage with the students at a personal level, or a combination of the two. A fifth theme commented upon was being able to work in groups, were allowed to talk and given some control their learning activities, and the resultant recognition that there are multiple ways to solve a problem. Each of these themes was viewed by students as contributing to their learning and appreciation of physics. In response to what contributes to their learning, one student, for example, said,

> There are lots of teachers that you have to do it their way. With my teacher it doesn't matter. But, he recognizes that everyone thinks differently and that as long as your way to get there is logical.

Students reported liking when their teachers use hands-on activities, a sixth theme. One student commented, "she does that hands-on thing before she starts to introduce us to what we will be doing. Real life things, like that's why she had the ball. So, she always does something with like a prop or something." The other students in the focus group agreed with her when she made the comment and gave further similar examples. In another group, a student said,

> I would say that the most beneficial part of this class has been the hands-on labs... because like [another student said] we haven't had much experience with that in the past but it is neat to see it in reality rather than in theory.

Themes three to six indicate students' enjoyed the classes when demonstrations and inquiry activities were used to engage them in the learning process.

Critical Context Barriers to Planning and Teaching Physics. Teachers noted students had been taught to regurgitate information in previous science courses and now had to think for themselves so, they were struggling. One teacher observed "I don't want students to regurgitate. I want students to be problem solvers." Another teacher said, "They don't know how to think outside the box and I try to get them to do so." Other difficulties associated with physics instruction included classroom interruptions. Multiple teachers mentioned the physics class period was the most interrupted class of the day. One younger teacher said, "Yeah, I never knew how many interruptions there are—assemblies, announcements—I mean I remember them in high school, but I didn't realize what a distraction they were." Another teacher confided she had recorded the number of interruptions her class received as an explanation to her administration for students' difficulties in understanding physics. Finally, in regard to hindrances to students' learning, some teachers and students said students' focus on grades was a distraction to their enjoyment of physics learning. One student said,

> Colleges will be looking at our grades. And, I think all of us would say it is important to us. But, it seems like no matter how hard we try, it is still hard for us to get an A average.

Students' Overall Perception of Their Physics Instruction. The majority of students interviewed in focus groups had a positive perception of their instructors and the way they taught their classes. Themes developed from student focus group interviews are summarized in Table 2.9. Students noted how they liked, as one said, "the fact that [the teacher] explains it thoroughly and goes back over it." Other students remarked they appreciated how their teacher, as one said, "gets to know each person," in an effort to tailor physics problems and tests to the students. In a number of focus groups, students, especially those who had the prerequisite mathematics knowledge, said they enjoyed their physics class as they were better able to understand the content and the world around them. Some focus groups agreed with the claim of one student that due to the instruction they received, they "became more confident as the year progressed."

Physics as a Path to a STEM Career. Most of the students enrolled in these physics courses were interested in pursuing a STEM career. Many students—especially those who liked physics, science in general, and mathematics—mentioned their desire for careers in engineering, the medical field, marine biology, nursing, physics, or dentistry. Students who did not want to pursue a STEM career

TABLE 2.9. Summary of Physics Student Interviews on Learning Physics

1.	Teacher's lack of knowledge (mathematics or physics) reported by students who were not able to grasp physics concepts because of poor mathematics preparation
2.	Teachers pacing of course concepts was very influential to students liking or disliking physics; student reported teachers who tried to meet all the standards seemed to make physics difficult to students
3.	Good physics teaching related to teacher confidence; higher confidence was noted by students
4.	Active and concrete learning was helpful as labs and demonstrations help students pay attention or engage in the material
5.	Interest increased with relevant applications as students enjoyed learning how physics applied to their everyday life
6.	Teacher dedication motivated students to enjoy physics primarily because of the teacher's interest and intensity
7.	Social aspects of classroom such as working in groups contributed to their learning of physics
8.	Creative approach to teaching providing encouragement in multiple ways to solve a problem contributed to appreciation of physics
9.	Critical barriers to learning physics included mathematics preparation, teacher quality and interest, and use of relevant inquiry teaching strategies

were typically students who reported not liking science or mathematics. Students' career choices were rarely influenced by their previous classes. Instead, in some cases, high school science classes were reported to merely serve as reinforcement for their interest in STEM careers. One student noted "I've been wanting to be a civil engineer since like 5th grade," while another said "Well, I was going to have a big thing like pediatric oncology and then I volunteered at a children's hospital in the summer and I fell in love with the idea of a being a nurse." While students indicated they assumed physics would be an important foundation for a STEM career, those not interested in such a career thought they should take physics as part of a basic college preparatory education.

Sub-Question 2: What teacher characteristics were related to the implementation of reform practice?

To answer sub-question 2, we construct profiles of four sample teachers in order to compare and contrast levels of reform practice. To compare and contrast the sample teachers at different times, four sample physics teacher profiles were selected based on a) Year 0 and Year 2 findings, b) RTOP ratings from observations, and c) years of teaching experience (see Table 2.10). Interview results support the profiles. A summary of interview results is found in Table 2.11 for Year 2.

Representative Teacher Profiles

Teacher 1
Over a 13 year teaching career, Teacher 1 had been teaching the AP physics course for four non-consecutive years at the time of the baseline, Year 0, observation. While she stated in her interview that helping students with "learning to think, asking questions, learn from errors, and become metacognitive" was the best way to teach physics, the observer reported her classroom was very teacher centered. An RTOP score of 11.5 supported the observer's conclusion as it was a low rating indicating there was very little evidence of reform in her AP physics

TABLE 2.10. Depicting Individual Teacher Interview Profiles

Teacher	PD Year	Level	RTOP	STEBI	LE	Certification	Years of Physics Teaching Experience
1	0	1	11.5	89	0.61	General	4
	2	3	51.5	82	0.73	Science	6
2	0	2	38.0	76	0.17	General	1
	2	4	82.0	86	0.87	Science	3
3	0	3	62.0	73	0.52	General	3
	2	4	88.5	99	0.69	Science	5
4	0	4	87.0	80	0.67	General	17
	2	4	83.5	100	0.95	Science	19

TABLE 2.11.   Summary of Year 2 Teacher Interview Themes on Teaching Physics

1.	Teachers repeatedly emphasized four topics: content emphasis, skills-focus, real-world relationships, and future orientation (academic and or career)
2.	Teachers utilize the APEX model approach to the extent it is possible and they find that it works well for them/their students
3.	Most teachers emphasize student-centered learning or inquiry in their classrooms
4.	Lecture seems to be a "dirty word," as teachers try to downplay their use of this method in the classroom
5.	Teachers expressed frustration with the following as barriers to using the APEX model: time issues, multiple preps, equipment and technology problems (access or function.
6.	Teachers indicated students' mathematics skills impeded their progress in the classroom, and that many students fear the mathematics portion of physics. While some mentioned the APEX approach has helped to make mathematics not as "in your face" for students, these fears/frustrations/limitations still exist
7.	A few teachers indicated they were uncomfortable or unfamiliar with the material being taught
8.	A few teachers lacked confidence that their students would understand the lesson taught. They reported that within in a few weeks, students' alternative misconceptions would come back after instruction.
9.	All teachers changed their teaching approach as a direct result of their participation in the APEX intervention professional development.
10.	Teachers increased their understanding of aspects of physics and mathematics content
11.	Teachers increased their understanding of physics inquiry teaching and learning
12.	Teachers increased their ability to implement differentiated instruction in a physics classroom
13.	Teachers decreased their isolation from other physics teachers
14.	Teachers increased their awareness of student prior knowledge and its use in lessons
15.	Teachers increased their ability to developing meaningful student engagement
16.	Teachers increased their ability to assess inquiry learning
17.	Teachers increased their understanding of teaching interpreted through physics classroom strategies
18.	Teachers increased their ability to find and use outside material and reference resources
19.	Physics was seen by students as a support course, not as major

classroom. Her student engagement rating, SLE, was 61%. During the student focus group interview, the students revealed that lecture, *PowerPoint* presentations, and homework problems were the primary instructional strategies used in the physics class. Students also indicated, as one said, the teacher "tries to do labs, once or twice a month." Although students stated the teaching practices employed during the observed lesson were effective, one student remarked he "would like to see real world applications of something... it helps your understanding." The teacher's self-reported efficacy score indicated she possessed high self-efficacy in her ability to teach science, as reflected in her PTEBI overall score of 89. When asked during the interview how confident she was about the lesson, she replied she was very confident noting she was confident that "students' ideas were reinforced." In her interview with the observer, she indicated she lacked physics content knowledge and looked forward to improving her content knowledge and pedagogical skills.

In Year 2's observation of Teacher 1, the observer noticed a change in classroom reform elements, as evidenced by an RTOP rating of 51.5, a moderate level of reform. After professional development, Teacher 1 reported, in Year 2, lecturing 30% of the time and having students work problems 40% of the time indicating she "[lectured] because they need the information." When asked if her teaching had changed she replied "Yes…I used to think that this is an advanced class so it is all about grades…[and] that…I almost don't care about the grade anymore…." The observer noted she did not see evidence of real world applications of the concepts taught, even though Teacher 1 stressed the importance of labs reinforcing the physics concept being studied in the class. Teacher 1 still reported a high efficacy score on the PTEBI, 82, remarking "APEX has helped tremendously on every aspect of it [teaching]. It has given me greater confidence in my skill." In their focus group interview, students reported the use of labs and practice problems as instructional tools, but did not report the use of *PowerPoint*s or lectures as had been reported in Year 0. One student even argued she could "have lectured a little," while another asked for "a softer introduction." Teacher 1 seemed to be engaging in more inquiry teaching in Year 2. The students were also more engaged during the follow-up, Year 2, classroom observation, with an SLE of 73% reported.

Teacher 2

In her ten years of teaching, Teacher 2 was teaching physics for the first time during the baseline Year 0 observation visit, his general physics class. He stated "I want my students to learn the various concepts related to physics content" and wished that "I had more available resources, so my lessons would be more efficient and more engaging." His teaching may have been influenced by these limitations because the observer only saw evidence of beginning levels of reform and reported a teacher centered classroom. Very little learning engagement was observed during the class. In their focus group interviews, students frequently stated they did not understand the concepts very well. One student wished there were "more labs and hands-on" during instruction, while another wished Teacher 2 "[interacted] with us more to Figure 2.out things we don't understand." They described a typical class as including notes, concept quizzes, practice problems, homework, going over homework, and taking notes." Teacher 2 reported a satisfactory level of self-efficacy in his ability to teach physics and was confident students "[would] understand the upcoming lesson" while in the focus group, one student noted she was still confused about "Friction…and all the concepts about it."

In Year 2, the teacher was recorded as evidencing a higher level of reform in physics instruction. The observer reported the teacher used APEX activities to teach the lesson, exhibited knowledge of the physics content, and made effective use of time. When asked about the best way to teach physics, Teacher 2 replied, "You do some lecture, keep it short, not very many notes. Work out some simple problems and just let them work in the classroom and kind of on their own." He also indicated his current students "were more interested in physics." During the focus group interview, students reported the class as interesting and enjoyable. Teacher 2 received a significantly higher student engagement rating in Year 2, with an SLE of 87%. He also reported a higher level of self-efficacy, PTEBI of 86%, in his ability to teach physics.

Teacher 3

At the time of her baseline Year 0 visit, Teacher 3 had been teaching physics for two years. In Year 0, she demonstrated moderate levels of reform, with an RTOP rating of 62. During her interview, she stated she "was not comfortable teaching physics the first two years but having attended several workshops, I am now more confident" and now "more comfortable with teaching physics concepts." The observer reported Teacher 3 was "very knowledgeable of the physics concepts taught during the observed lesson." Teacher 3 reported a satisfactory level of self-efficacy in her ability to teach physics. Her PTEBI score was 62. She believed "instruction should facilitate meaningful learning and that is accomplished through lecture, hands-on activities, and student discussion." As lecture was an integral and prominent part of the instructional strategies, a possible need for more reform in the classroom was indicated. Her student engagement levels during the lessons were low, 52%. The focus group students, who all reported an interest in STEM careers, enjoyed their physics class. They agreed with one student's comment that they "like hands-on activities because they seem practical and relevant to our lives and to [physics class] learning." One student further stated he saw "how the concepts on velocity, displacement, acceleration, and distance aid my future career."

In Year 2, Teacher 3 displayed a higher level of reform, as indicated by an RTOP rating of 88.5. She attributed change in her instructional practices to the APEX intervention professional development. During her interview she stated, "You know…and I was always one to do a lot of labs and a lot of projects and stuff and this was one I did even before APEX…" She continued "but, one of the things I have found with APEX is by getting them involved in the assignment, collecting the data, doing that four-step analysis which is the best thing ever, ummm…." She reported a higher level of self-efficacy in her ability to teach physics, accomplishing a PTEBI score of 99. In the focus group interview, students often noted hands-on activities as most beneficial to their learning. Students were engaged most of the time during the lesson, with an SLE of 69%. Although only one of the six students interviewed was interested in a STEM career, the group members enjoyed the class, thought it beneficial to them, and appreciated that they had, as one said, "more opportunities to learn through lab" rather the memorization of "a lot of information" which characterized their other science classes.

Teacher 4

In his 17th year of teaching physics, Teacher 4 proved to be exemplary in engaging in teaching practices rooted in reform. His Year 0 RTOP rating was 87. When asked what was the best method for teaching physics he replied "Socratic questioning, direct instruction, group work, cooperative learning—I guess you throw all those things together, and that's my method." He believed his pedagogical approach proved effective because he had "gotten emails, notes, letters, from students in college telling me that—I can show you some—'I got an A in physics, thank you.' Or, 'My physics teacher stinks, and I'm using your notebook, thank you.'" The observer reported classroom dynamics such as the teacher walking around the classroom, the class members being relaxed, and the teacher considering and addressing all student questions.

Teacher 4 strongly emphasized real-life applications which the observer noted as meaningful to the students. The observer did, however, notice some students were off task during group work in the lab and, despite his high RTOP rating, Teacher 4 received a median engagement rating with an SLE of 67%. Students raved about Teacher 4 during the focus group discussion. They responded, "this is an interesting class," "he is thorough" and "I really like him" when talking about their physics class. Students revealed evidence of inquiry teaching. One student stated "There's no baby steps, there's just, here's your problem. Do it." Another student noted what he learned in his physics class "... helps in other classes too." Teacher 4 had self-efficacy in his ability to teach physics, with a PTEBI score of 80.

In Year 2, Teacher 4 continued to display evidence of reform in his instructional practices. He had not adapted many additional APEX teaching strategies, except white boarding, partly because he already incorporated a number of them as had been observed during the Year 0 visit. Teacher 4 had settled on a way of teaching that he considered as successful for him, meaning students performed well compared to national standards, so he did not believe in the need to move toward more open-ended inquiry. He preferred directed inquiry labs, but his terminology for directed inquiry and open-inquiry differed slightly from how these were described in the APEX model. For him, directed inquiry had open-ended inquiry elements; it was not merely verification. Teacher 4 reported a higher level of self-efficacy about his ability to teach physics in Year 2. His PTEBI score was 95. He stated, "So, I know my method works. I have the data to back that up." The data here being "last year their scores were very good. They scored higher than the national average and higher than the global average which is pretty good."

Teacher 4's students seemed engaged and needed little direction. He had a much higher student engagement in Year 2 with a rating on the SLE of 95%. The students almost seemed self-sufficient to the observer, hardly ever drawing on the teacher's guidance through the lab. The students reported enthusiasm for teacher

4's method of instruction and felt he was their favorite teacher. One student described her feelings about the class.

> Very positively actually—I didn't really enjoy science until I got to chemistry and physics and this physics class in particular—I am more confident about science going into college and I'm more confident about science—my ability to think things through in a way that's not just step-by-step: reason my way through more. Scientific things. Very, very positively.

Although taking Teacher 4's class seemed to be challenging for them, the students reported increased self-confidence in their problem-solving ability as a result of the class in addition to an overall greater appreciation for science and their ability to "do science." One student stated "[Teacher 4] prepared us well for our college physics classes. I am not as intimidated going into next year as I would be if I had didn't have him."

Teacher Efficacy

Teacher characteristics were investigated to determine which were related to the implementation of physics reform practice in their classrooms. Several variables were investigated. The *Physics Teaching Efficacy Belief Instrument* (PTEBI) adapted from the STEBI-A (Riggs & Enochs, 1990) assessed in-service science teachers' personal self-efficacy belief. Efficacy was a teacher characteristic measured before and during the professional development intervention program. A one-way ANOVA revealed no significant differences in the total PTEBI scores for Year 0 and Year 2; $F(1, 80) = 1.59$, $\alpha = .21$. Teachers did not rate themselves significantly differently after receiving professional development on the instrument total score (see Table 2.12). PTEBI-A has two distinct subscales. The subscale considered first was personal physics teaching efficacy (PSTE) which measures teachers' beliefs in their own ability to teach physics. Investigating results on the PSTE subscale, a one-way ANOVA revealed no significant difference in the scores for Year 0 and Year 2; $F(1, 80) = 1.22$, $\alpha = .27$. Teachers did not rate their beliefs in their own ability to teach science differently after receiving professional development. The PTOE subscale assesses teachers' beliefs that student learning can be influenced by effective teaching. For PTOE a one-way ANOVA revealed no significant difference in the scores for Year 0 and Year 2; $F(1, 80) = 3.62$, $\alpha = .061$. Teachers' beliefs about their ability to influence student outcomes did not change after receiving professional development.

TABLE 2.12. Summary of Teachers' Perceptions of Efficacy using PTEBI

Year	Total Mean (SD)	Efficacy beliefs Mean (SD)	Outcome expectancy Mean (SD)
0	78.54 (10.47)	42.41 (6.63)	36.12 (5.84)
2	81.90 (13.53)	43.98 (6.16)	38.73 (6.42)

Sub-Question 3: What effect did reformed classroom practices have on student learning?

Physics students took pre- and post-tests on unit topics where teachers' had received intervention training in physics content knowledge and appropriate pedagogy in the APEX model professional development program. One classroom unit investigated in Year 2, for example, was on Force and Motion and Newton's Laws. A modified version of the standardized *Force Concept Inventory* (FCI) (Hestines, Wells, & Swackhamer, 1992) was used for testing. Students FCI pre-test mean score (class mean) was 25.8% and their mean post-test score was 47.6% of the total of 100%. An ANOVA comparison of means, $F(1,753) = 197.96$ at $p < 0.01$, was significant with an Effect Size of 1.03. Each physics class FCI mean gain score was compared to the RTOP rating of reform in their classroom and years of science teaching experience. A significant predictor of FCI achievement gain score $F = 26.80$ at $p < 0.01$ was the RTOP rating of the teacher: students whose teacher fell into the moderate or high RTOP rating of classroom reform showed better performance on the FCI. The single classroom reform regression measure explained 23% of the variation in gain scores, $R^2 = 23\%$, see Figure 2.4.

The impact of a teacher's level of reform based on their RTOP rating was related to a student class FCI normalized mean gain score. A Traditional or Beginning reform level, RTOP of 50 or less, was related to an FCI gain score of 5–15% on the pre- posttest during the Force and Motion unit. A Moderate or High reform

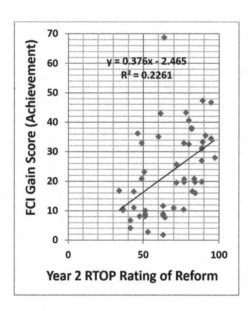

FIGURE 2.4. Classroom Reform Level Compared to Achievement

level, RTOP of 50 or more, was related to an average FCI gain score of 25–35% on the pre-posttest during the Force and Motion unit (Figure 2.4).

Multiple linear regression analyses were conducted to predict gain scores with type of physics class taught, number of APEX strategies used in instruction and RTOP rating. In the method of dummy coding employed in this analysis, indicator coding, data for two of the explanatory variables (class type and RTOP level) were transformed into dichotomies and coded as absent (0) or present (1) in that category. The continuous variable of APEX strategies was entered in the first step to predict total gain score (Model 1), class type and RTOP classification were then simultaneously added (Models 2 and 3, respectively). All the three models significantly predicted gain scores, while accounting for a significant proportion of variance (Model 1: $F_{1, 30}$, = 8.008, $p < .01$, $R^2 = .21$; Model 2: $F_{5, 26}$, = 3.686, $p < .05$, $R^2 = .415$; Model 3: $F_{8, 23}$, = 3.392, $p < .05$, $R^2 = .541$). See Table 2.13.

Sub-Question 4: How has the classroom learning environment changed?

During the first and second year of the intervention professional development model program, teachers conducted and submitted classroom action research reports related to their common annual Force and Motion unit. The unit description report, including lesson plans, was analyzed for types of teaching strategies used (see Table 2.14). The extent of implementation of those APEX PD strategies experienced during the intervention's professional development program was defined as a fidelity rating of reform. A total of 18 teaching strategies were identified from the action research report lesson plans. Eight APEX model teaching strategies were found to be common in most classroom lessons where students accomplished higher than average normalized mean gain scores on the FCI test, see Table 2.14. An ANOVA comparison of means between the number and type of APEX strategies found in lesson plans, the fidelity rating, and students' FCI mean

TABLE 2.13. Individual and Group Differences in Gain Scores

	FCI Gain Scores	
	M	**SD**
Type of Physics Class		
Regular	20.11	17.07
Honors	16.22	12.11
AP Physics	30.51	18.08
Level of Reform (RTOP Level)		
Traditional	16.53	14.66
Beginning	19.95	14.61
Moderate	18.21	11.94
High	29.09	18.40

TABLE 2.14. Teacher Selected PD Instructional Strategies Related to Higher Student Achievement

Common PD performance indicators* of successful teaching, during classroom force and motion units, found to be similar in classes where students scored above the FCI gain mean.

1. Graphical analysis of data in a 4 step analysis with mathematical modeling

2. Guided inquiry laboratory activities

3. Identification and use of student alternative conceptions

4. Public presentations and argumentation with students explaining and defending results

5. Use of technology to facilitate student learning and student use (as opposed to teacher's use in class).

6. APEX, AAPT/PTRA and other lesson materials were used in planning lessons (AAPT/PTRA - American Association of Physics Teachers and Physics Teaching Resource Agent)

7. Free body diagrams used in student labs

8. Increased student talk and control of learning activities during lessons

*rated by outside reviewer

gain scores, $F=8.00$ at $p=0.02$, was significant (see Figure 2.5). For example, a physics teacher's use of only two APEX model strategies in reported lessons was related to a student class FCI mean gain score of less than 10% on the pre- and post-tests during the Force and Motion unit. Classes using six or more reported lesson strategies had class FCI mean gain scores of 20–40% (see Figure 2.5).

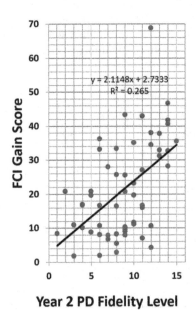

Year 2 PD Fidelity Level

FIGURE 2.5. Classroom Professional Development Fidelity Level Compared to Achievement

As an additional check on empirical validity of the instruments, teacher RTOP rating was compared to the APEX model lesson fidelity rating. A significant relationship between teachers' RTOP and APEX model lesson fidelity was found using an ANOVA comparison of means, $F=6.97$ at $p=0.046$. A physics teacher's RTOP rating of 36, for example, was related to the teacher's use of two APEX model lesson strategies while a rating of 85 related to a teacher's use of six or more APEX model strategies in lesson plans.

The number of APEX strategies used in instruction was a significant predictor of variance in gain scores (see Table 2.15). Students whose teachers used APEX strategies were 2.56 times more likely to achieve higher gain scores. Class type was also a significant predictor of variance in gain scores: students in AP physics classes obtained gain scores of 14.62 intervals higher than students in honors physics classes ($p = .02$) and 13.87 intervals higher than students in general phys-

TABLE 2.15. Regression Analyses for Model Items With Dummy Coded Categorical Variables

	β	SE	t	p
Model 1: FCI Gain Scores				
Constant	21.531	2.529	8.515	< .01
APEX PD Fidelity	2.556	.903	2.830	< .01
Model 2: FCI Gain Scores				
Constant[a]	29.618	4.186	7.076	<.01
APEX PD Fidelity	2.694	.845	3.190	<.01
Regular	−2.194	7.293	−.301	.766
Honors	−14.615	6.069	−2.408	.02
Pre-AP	−19.967	13.852	−1.441	.16
General	−13.867	6.509	−2.130	.04
Model 3: FCI Gain Scores				
Constant[a]	37.753	5.563	6.787	< .01
APEX PD Fidelity	2.924	.828	3.530	< .01
Regular	−7.700	7.420	−1.038	.31
Honors	−17.555	5.934	−2.958	< .01
Pre-AP	−16.913	14.004	−1.208	.24
General	−12.515	6.982	−1.793	.09
Traditional	−18.059	7.535	−2.397	.03
Beginning	−6.130	5.926	−1.034	.31
Moderate	−11.241	7.922	−1.419	.17

[a] In the regression equation, the constant values represents the dummy coded comparison categories for: AP Physics, and High RTOP level

ics classes ($p = .04$). Table 2.13 reports students whose teachers demonstrated high levels of reformed physics instruction obtained significantly higher gain scores of 18.06 intervals than students who were taught by a traditional teacher ($p = .03$). The regression analysis indicated variance in gain scores could be predicted from increased use of APEX instructional strategies, class type, and RTOP classification. Together these variables account for 54.1% of the variance in gain scores.

CROSS-STRAND INTEGRATION AND CONCLUSION

Significant results were found with and between three sources of data; 1) outside observer's quantitative and qualitative reports of a physics teacher's reform, 2) student achievement, and 3) fidelity of teachers' APEX model implementation. Triangulation of results indicated physics focused professional development fostered significant differences in the way physics teachers structured classrooms, conducted teaching, and engaged students with greater student learning as a result. Similar reform findings were supported in each of three separate parallel studies using the convergent parallel design.

Critical differences in teaching approaches were found in teacher interviews between pre- and post-intervention professional development. Before the professional development began, in Year 0, sample teachers extensively reported traditional lessons focused on lecture and recitation, with a few hands-on activities and practices in some of their lessons. This method was generally described by teachers as "inquiry teaching" in their physics classrooms.

Biology was the academic major of the great majority, over 80%, of the sample teachers. Their classroom activity focus approached, but did not reach, the inquiry emphasis of the process of investigation used in the biological sciences. The hands-on emphasis in lessons by exemplary biology teachers is on helping students to develop their own questions or generate a hypothesis. This emphasis leads to the development of experiments through research questions thus allowing students to understand and communicate cause-effect relationships. Reformed biology practice differs from that of reformed chemistry or physics teachers (Breslyn & McGinnis, 2012).

Following the APEX intervention professional development, in Year 2 sample teachers reported in interviews and via their lesson plans that inquiry physics teaching included using mathematical modelling to develop physics relationships rather than experiments to verify concepts and relationships. Modeling often involved developing mathematical equations to describe and predict physical phenomena. In modeling, students are presented with a problem, decide what data to collect, and decide what procedure to use to report the data. On the basis of the data, students conduct an analysis involving graphing to generate a mathematical model in the form of an equation to describe the phenomenon, relationship between variables, and predict its behavior. Exemplary physics teachers report physics inquiry as involving searching for patterns and relationships culminating in modeling, predicting outcomes, and determining the best data explanations, equations, simulations, or models (Breslyn & McGinnis, 2012). Our study's interviews, rather than being indicative of the sample teachers' overwhelmingly traditional biology teaching approach, were now, in Year 2, documenting efforts to follow the patterns researchers have identified as found among exemplary teachers of physics.

A complementary study, following the study reported in this chapter, supported the results (Ogodo, 2017). This case study examined how participation in additional APEX physics focused professional development influenced the instructional practice of AP physics teachers for a small sample selected from a statewide population of physics teachers. Fourteen AP teachers who received College Board AP training were purposefully selected from high schools in the state of Alabama. From this population, seven teachers who participated in additional APEX physics-focused professional development beyond the College Board training (APEX PD) were compared with the other seven teachers who had little or no additional in-service physics focused training (AP-only PD). The average reform, RTOP, rating for the APEX PD teacher group pre-intervention was 50.2 and for the AP-Only PD teacher group 54.7. After the APEX intervention, the APEX PD teachers were rated higher than the AP-only PD teachers. The mean rating for the APEX PD teachers increased from 50.2 to 77.1 (Ogodo, 2017). Analysis of the quantitative data from the RTOP classroom observation indicated a large treatment effect difference between the APEX PD teachers and the AP-only PD teachers. The result, $t(12) = .32$, $p = .008$ found the APEX PD teachers' mean RTOP ratings were significantly larger than those for the AP-only PD teachers. APEX PD teachers implemented more reformed-based practices in their classrooms and used less teacher-centered instruction compared to the AP-only PD teachers' average rating. A large effect size difference of $d = 1.15$ was found between the groups based on Cohen's guidelines (Ferguson, 2009). The qualitative data from teacher interviews, surveys, and questionnaires were thematically analyzed indicating the APEX PD teachers increased their PCK level after the intervention. In addition, the study found that although contextual factors influenced the type of instruction used by the teachers, these factors were more teacher-based than context-based.

The results provide a unique picture of the variables confronting physics teachers and have implications for in-service physics teachers' professional development. The implications have impact on how the professional development needs of in-service physics teachers can be addressed. The average physics teacher needs continuous long-term opportunities to develop integrated physics content knowledge and pedagogical content knowledge before effective reform is operationalized in the classroom. Physics is a major component of high school science, yet the number of well-prepared physics teachers is small. Professional development facilitating effective physics classroom reform is needed. Our study describes a professional development model and research strategies that, if implemented elsewhere, can work to advance the quality of physics education in the United States.

ACKNOWLEGEMENT

This material is based upon work supported by the National Science Foundation under MSP grant award DUE-1238192. Any opinions, findings, and conclusions or recommendations expressed in this material are those of the authors and do not necessarily reflect the views of the National Science Foundation.

REFERENCES

Abell, S., & Lederman, N. (2007). *Handbook of research on science education.* Mahwah, NJ: Lawrence Erlbaum.

Banilower, E. (2013). *2012 National survey of science and mathematics education.* Chapel Hill, NC: Horizon Research, Inc.

Breakspear, S. (2012). *The policy impact of PISA: An exploration of the normative effects of international benchmarking in school system performance.* OECD Education Working Papers, (71), Paris, FR: OECD Publishing. http://dx.doi.org/10.1787/5k9fdfqffr28-en

Breakspear, S. (2017). *Developing agile leaders of learning.* Wise Learn/Labs. Retrieved from file:///D:/25%20NOYCE%20TRACK%202/Article%202017%20Breakspear%20Developing%20Agile%20Leaders.pdf

Breslyn, W., & McGinnis, R. (2012). A comparison of exemplary biology, chemistry, earth science, and physics teachers' conceptions and enactment of inquiry. *Science Education, 96,* 48–77.

Budd, D. A., Van der Hoeven Kraft, K. J., McConnell, D. A., & Vislova, T. (2013). Characterizing teaching in introductory geology courses: Measuring classroom practices. *Journal of Geoscience Education, 61*(4), 461–475.

Creswell, J., & Plano-Clark, V. (2011). *Designing and conducting mixed methods research.* Thousand Oaks, CA: Sage.

Crouch, C., Watkins, J., Fagen, A., & Mazur, E. (2011). Peer instruction: Engaging students one-on-one, all at once. In E. Redish & P. Cooney (Eds.), *Research-based reform of university physics.* College Park: MD: American Association of Physics Teachers, Reviews in PER Vol. 1, Reviews in PER Vol. 1, http://www.per-central.org/document/ServeFile.cfm?ID=4990.

Crouch, C., & Mazur, E. (2001). Peer instruction: Ten years of experience and results. *American Journal of Physics, 69,* 970–977.

Desimone, L. M. (2011). A primer on effective professional development. *Phi Delta Kappan, 92*(6), 68–71.

Desimone, L., Porter, A.C., Garet, M., Yoon, K. S., & Birman, B. (2002). Effects of professional development on teachers' instruction: Results from a three-year study. *Educational Evaluation and Policy Analysis, 24*(2), 81–112.

Druit, R. S. (2014). Teaching physics. In N. G. Lederman & S. K. Abell (Eds.), *Handbook of research on science education* (pp. 434–451). New York, NY: Routledge.

Ferguson, C. J. (2009). An effect size primer: A guide for clinicians and researchers. *Professional Psychology: Research and Practice, 40*(5), 532–538.

Gray, L., & Taie, S. (2015). *Public school teacher attrition and mobility in the first five years: Results from the first through fifth waves of the 2007–08 beginning teacher longitudinal study. First Look* (NCES 2015-337). Washington, DC: National Center for Education Statistics.

Hestenes, D., Wells, M., & Swackhamer, G. (1992). Force concept inventory. *The Physics Teacher,(30,* 141–151.

Hodapp, T., Hehn, J., & Hein, W. (2009) Preparing high-school physics teachers. *Physics Today, 62*(2), 40–45.

Knight, R. D. (2004). *Five easy lessons: Strategies for successful physics teaching.* Boston, MA: Addison Wesley.

Lederman, N., & Gess-Newsome, J. (1999). Reconceptualizing secondary science teacher education. In J. Gess-Newsome & N. G. Lederman (Eds.), *Examining pedagogical content knowledge* (pp. 199–214). Boston, MA: Kluwer.

Loughran, J., Berry, A., & Mulhall, P. (2012). *Portraying PCK. Understanding and developing science teachers' pedagogical content knowledge* (pp. 15–23). Rotterdam, Netherlands: Sense Publishers.

Loughran, J., Mulhall, P., & Berry, A. (2004). In search of pedagogical knowledge in science: developing ways of articulating and documenting professional practice. *Journal of Research in Science Teaching. 41*(4), 370–391.

MacIsaac, D., & Falconer, K. (2002). Reforming physics instruction via RTOP. *The Physics Teacher, 40*(November), 16–21.

Meltzer, D. E., & Otero, V. K. (2015). A brief history of physics education in the United States. *American Journal of Physics, 83*(5), 447–458.

National Research Council. (1996). *National science education standards.* Washington, DC: National Academies Press.

National Research Council. (2005). How students learn: Science in the classroom. In M. S. Donovan & J.D. Bransford (Eds.), *A targeted resource for teachers.* Washington, DC: Committee on How People Learn, National Academy Press.

National Research Council. (2013). *Next generation science standards* (NGSS) Washington DC: National Academy Press. http://www.nextgenscience.org/

O'Brien, G., & McIntyre, A. (2011). *Designing effective teacher professional learning for improved student outcomes—Research findings from NSW schools.* Paper prepared for the ACEL 2011 Annual Conference—Learning Landscapes: Strategies for Sustaining Change, Adelaide, Australia.

Ogodo, J. (2017). *Influence of physics focused professional development on advanced placement physics teachers.* (Unpublished doctoral dissertation). University of Alabama, Tuscaloosa.

Paek, P. L., Ponte, E., Sigel, I., Braun, H., & Powers, D. (2005). A portrait of advanced placement teachers' practices. *ETS Research Report Series, 2005*(1), i–41.

Riggs, I. M., & Enochs. L. G. (1990). Toward the development of an elementary teacher's science teaching efficacy belief instrument. *Science Education, 74*(6), 625–637.

Rock, J., Courtney, R., & Handwerk, P. (2009). *Supplementing a traditional math curriculum with an inquiry-based curriculum: A Pilot of math out of the box.* Princeton, NJ: Educational Testing Service.

Saunders, R. (2014). Effectiveness of research-based teacher professional development. *Australian Journal of Teacher Education, 39*(4), 166–184.

Sawada, D., & Piburn, M. (2000). *Reformed teaching observation protocol (RTOP).* (ACEPT Technical Report No. IN00-1). Tempe, AZ: Arizona Collaborative for Excellence in the Preparation of Teachers.

Sawada, D., Turley, J., Falconer, K., Benford, R., & Bloom, I. (2002). Measuring reform practices in science and mathematics classrooms: the reformed teaching observation protocol. *School Science and Mathematics, 102*(6), 245–252.

Seastrom, M. G. (2004). *Qualifications of the public-school workforce: Prevalence of out-of-field teaching.* U. S. Department of Education. Washington, DC: National Center for Educational Statistics. Retrieved from http://nces.ed.gov/pubs2002/2002603.pdf

Shulman, L. (1986). Those who understand: Knowledge growth in teaching. *Educational Researcher, 57*(1), 4–14.

Shulman, L. (1987). Knowledge and teaching: Foundations of the new reform. *Harvard Educational Review, 57*(1), 1–23.

Smith, D. C. (1999). Changing our teaching: The role of pedagogical content knowledge in elementary science. In J. Gess-Newsome & N. G. Lederman (Eds.), *Examining pedagogical content knowledge* (pp. 163–198). Boston, MA: Kluwer.

Sunal, D. (2013). *UA APEX five year research plan of work—Data collection protocol. 2013 annual report: National Science Foundation.* Tuscaloosa AL: The University of Alabama.

Sunal, D., Dantzler, J., Sunal C., Turner, D. Harrell, J. W., Aggarwal, M., & Simon, M. (2016). The 21st century physics classroom: What students, teachers, and classroom observers report. *School Science and Mathematics. 116*(3) 116–126.

Sunal, D., Sunal, C., Turner, D., Steele, E., Mason, C., Lardy, C., Zollman, D., Matloob-Haghanikar, M., & Murphy, S. (2014). National study of education in undergraduate science: research design. In Sunal, D., Sunal, C., Wright, E., Mason, C., & Zollman, D. (Eds.), *Research based undergraduate science teaching* (pp. 35–66). Charlotte, NC: Information Age Pub.

Thissen-Roe, A., Hunt, E., & Minstrell, J. (2004). The DIAGNOSER project: Combining assessment and learning. *Behavior Research Methods, Instruments, & Computers 36*(2), 234. https://doi.org/10.3758/BF03195568

Turner, D., & Sunal. bD. (2014). Investigating the long-term impact of undergraduate science reform courses on the pedagogical practices of kindergarten through sixth grade elementary teachers. In Sunal, D., Sunal, C., Wright, E., Mason, C., & Zollman, D. (Eds.), *Research based undergraduate science teaching* (pp. 121–152). Charlotte, NC: Information Age.

Van Driel, J. H., & Berry, A. (2012). Teacher professional development focusing on pedagogical content knowledge. *Educational Researcher, 41*(1), 26–28.

Van Duzor, A. G. (2011). Capitalizing on teacher expertise: Motivations for contemplating transfer from professional development to the classroom. *Journal of Science Education and Technology, 20*(4), 363–374.

White, S., & Tesfaye, C. (2010). High school physics courses and enrollments-Results from the 2008-09 nationwide survey of high school physics teachers. *Focus On,* August. American Institute of Physics Statistical Research Center. Retrieved from http://www.aip.org/statistics/trends/reports/highschool3.pdf.

White, S. & Tyler, J. (2014). *Who teaches high school physics? Results from the 2012–2013 nationwide survey of high school physics teachers. Focus On,* December. American Institute of Physics Statistical Research Center. Retrieved from http://www.aip.org/statistics/trends/reports/highschool3.professional developmentf.

CHAPTER 3

DESTABILIZING THE STATUS QUO IN STEM PROFESSIONAL DEVELOPMENT WITH MODELING INSTRUCTION

Kathleen A. Harper, Ted M. Clark, and Lin Ding

Modeling instruction (MI) is an approach used in a large number of high school science courses. Thousands of teachers have learned MI at professional development workshops and have rated the experience highly. However, an analysis of this training vis-à-vis characteristics of effective professional development has not been reported. To fill this gap, a three-week physics MI workshop spanning 12 years of operation has been studied to investigate how MI activities exemplify effective PD. Focus is on the participants' views and attitudes toward their experiences, the alignment of such with the program goals, and how the workshop format leads them to implement MI in their classes. Direct observation of workshop activities, along with quantitative and qualitative surveys completed by participants at the start of, during, and after the workshop, then after a school year, were examined across the 12 years. Findings support the prevailing view that effective PD addresses issues of content learning, coherence, participation, active learning, and networking, given sufficient duration to include these elements. They also indicate that activities such as having teachers role play as students in lessons led by high school teachers trained in MI are successful in educating teachers and influencing their practices.

Physics Teaching and Learning: Challenging the Paradigm, pages 57–83.
Copyright © 2019 by Information Age Publishing
All rights of reproduction in any form reserved.

Keywords: professional development, modeling instruction, STEM education, high school, active learning, pedagogical content knowledge

In my 23 years of teaching and engaging professional development for teachers at a host of institutional sites and theoretical emphases, the Modeling Workshop is the best, most useful in direct application to the classroom. The encouragement and affective support for teachers, many of whom come with rather frustrating classroom experiences makes a profound difference. A learning community is built and supported (I still have frequent contact with several teachers from the original workshop I attended). The commitment of workshop leaders, their patience to interact with participants, to clarify difficult concepts and offer encouragement and advice is of such high value. There is no professional development experience that I would endorse with as much enthusiasm as the Modeling Workshop. (Upton, 2014, p. 29)

—Modeling workshop participant

Unpacking this epigraph reveals a teacher's perspective of successful professional development (PD). For kindergarten–12 teachers, PD is perceived to be a key resource in the ongoing effort to raise student academic achievement, and this teacher strongly advocates for a particular approach, a modeling workshop, in comparison to other PD activities in which he has participated. Various reasons underpin this participant's enthusiasm, including the alignment with his own instructional goals ("direct application to the class"), a sense of community, and the contribution of workshop leaders. Implicit in this epigraph is the notion that even veteran teachers can benefit from quality PD. What makes a PD program beneficial for participants? What features motivate participants to adopt new practices? How does modeling instruction exemplify high-quality PD? As contributors to PD workshops, we find these questions profound and their answers informative. The questions framed an investigation of a particular well-established modeling instruction workshop during its first 12 years of operation. We begin with a review of literature on teacher PD, describe modeling instruction and its associated PD, then analyze the relationship between recommendations of the literature and elements of the modeling workshop through the lens of participant feedback.

LITERATURE REVIEW

Teacher Professional Development

The notion that a teacher's classroom practices influence the performance of students is not controversial, along with the idea that educators, and hence students, will benefit from on-going training that affords them the opportunity to learn about and implement educational best practices. This consensus has led to an enormous investment in the professional development of teachers, as conventional wisdom holds that we are on the verge of dramatic gains if we can provide teachers with the appropriate support and training (Jacob & McGovern, 2015). How large of an investment? It has been estimated that, after accounting for teacher time, personnel and staff support, and substitute coverage, districts spend an average of $18,000 per teacher each year on teacher development (Jacob & McGovern, p. 2). Is this too much? As noted in an overview examining this cost, "[a]n outsized investment in teacher improvement is not necessarily unwise or unmerited. The problem is our indifference to its impact—that all this help doesn't appear to be helping all that much" (Jacob & McGovern, p. 10).

Although various policy documents like the USA's No Child Left Behind Act of 2001 may require states to ensure the availability of high quality PD for all teachers, they do not address questions as to what constitutes high quality PD or how it should be made available (Borko, 2004). Recognition by policy makers that PD is important has not necessarily led to improvements in its content or format, leading to its characterization as "woefully inadequate," as billions of dollars are spent on in-service seminars and other forms of PD that are "fragmented, intellectually superficial and do not take into account what we know about how teachers learn." (Borko, p. 3).

A decade ago it was proposed that researchers were only just beginning to learn what and how teachers learn from PD and how such efforts impact student outcomes (Borko, 2009). It was hoped that, with greater insights, professional development policy and practice would improve. As Jacob suggests, however, it is far from clear that PD has changed significantly, or that the outcomes are now favorable. This is not to say that research has failed to provide insights into effective PD practices. On the contrary, a consensus is emerging on what makes PD effective (Birman, Desimone, Porter, & Garet, 2000; Garet, Porter, Desimone, Birman, & Yoon, 2001; Reimers, Farmer, & Klein-Gardner, 2015; Timperley, 2011; van Driel, Beijaard, & Verloop, 2001). A significant gap exists between the evidence-based best practices commonly espoused for quality PD, and the actual practices most commonly found. The paradigm for effective PD is all too often an academic one that does not guide or inform current practices.

Our investigation does not challenge the existing paradigm of quality PD. Rather, it supports this paradigm and challenges the PD status quo that diverges from these suggested practices, often resulting in little impact upon the profession. Our contribution has transformative potential in the sense that it provides an exemplar of evidence-based PD that may inform other reform efforts. This is not to claim our work discusses the only successful PD approach; PD is complex and not one-size-fits-all. We claim, however, that the reported guiding principles for PD do seem to have validity as demonstrated by the success of a program aligned with these notions.

ELEMENTS OF SUCCESSFUL PROFESSIONAL DEVELOPMENT

Six elements reported for effective PD in the literature are duration, content, active learning, participation, type of activity, and coherence. We describe these below and will use them as reference points in illustrating the effectiveness of modeling workshops and in interpreting the feedback provided by the participants.

Duration

Research indicates that longer duration PD activities are more likely to provide an opportunity for teachers to engage in deep discussion of content, of student conceptions and misconceptions, and of pedagogical strategies. Activities extending over time are more likely to allow teachers to try new classroom practices and obtain feedback (Garet et al, 2001). Such activities also create a stronger impact on teaching practice and are aligned more closely with systemic reform (Cobb, Wood, Yackel, Nicholls, Wheatley, Trigatti, & Perlwitz, 1991; Corcoran, 1995; Darling-Hammond, 1995; Hargreaves & Fullan, 1992; Hiebert, 1999; Lieberman, 1996; Little, 1993; Richardson, 1994; Sparks & Loucks-Horsley 1989; Stiles, Loucks-Horsley, & Hewson, 1996; Supovitz & Turner, 2000; Wood & Sellers, 1997).

Content

The degree to which PD focuses on content knowledge is directly related to teachers' reported increase in knowledge and skills. Teachers do not find it effective if the generic PD only focuses on teaching techniques without emphasizing content (Birman et al., 2000; Cohen & Hill, 1998; Garet, et al., 2001; Kennedy, 1998). PD focusing on mathematics and science content and the ways students learn such content has been widely found to be helpful, particularly in improving students' conceptual understanding (Cohen & Hill; Fennema, Carpenter, Franke, Levi, Jacobs, & Empson, 1996).

Active Learning

Teachers who participate in PD with active learning opportunities report increased knowledge and skills, as well as improved classroom practices (Birman et al., 2000). Active learning encourages teachers to engage in meaningful discussion, planning, and practice (Garet et al., 2001; Lieberman, 1996; Loucks-Horsley, Hewson, Love, & Stiles, 1998). Such learning can take various forms, including observing expert teachers, having one's own teaching observed, planning how new curriculum materials and teaching methods can be used in the classroom, reviewing student work in topic areas, and leading student discussions (Garet et al.).

Participation

Collective participation is yet another important factor that positively contributes to effective PD (Birman et al., 2000). Such participation enables teachers to discuss concepts and problems arising from the professional development activities. It also allows participants to integrate what they have learned with other aspects of instructional content by establishing a shared professional culture. Within this culture, teachers in the same school or teachers who teach the same content can develop a common understanding of instructional goals, methods, problems, and solutions (Ball, 1996; Garet et al., 2001; Newmann & Associates, 1996).

Type of Activity

Traditional approaches, as described below, are less effective than reformed approaches. Reformed approaches include study groups, teacher networks, mentoring relationships, committees or task forces, internships, individual research projects, teacher resource centers, and peer coaching (Birman et al., 2000; Garet et al., 2001). Traditional formats are criticized for not giving teachers the time, activities, and content necessary to increase knowledge and foster meaningful change in classroom practice (Loucks-Horsley, Hewson, Love, & Stiles, 1998). Reformed PD activities, in contrast, are more responsive to how teachers learn (Ball, 1996) and have a greater influence on their teaching practices (Darling-Hammond, 1995; Hargreaves & Fullan, 1992; Little, 1993; Richardson, 1994; Sparks & Loucks-Horsley, 1989; Stiles, Loucks-Horsley, & Hewson, 1996). In particular, learning networks, peer coaching, collaborative action research, and case studies have powerful potential for lasting changes in teachers' practical knowledge (van Driel et al., 2001).

Coherence

Coherence of PD directly impacts teachers' learning. Coherent activities align with teachers' goals, draw on their earlier experiences, and encourage them to discuss their experiences with other teachers and administrators. Equally importantly, coherent PD aligns closely with national, state and district standards and assessments (Birman et al., 2000).

Traditional Professional Development Programs

Traditional professional development diverges from these recommended practices. Most PD is short in duration, with teachers reporting activities lasting a total of one to eight hours during each 12-month calendar year (Birman et al., 2000). Moreover, traditional PD primarily focuses on general principles, teaching techniques, beliefs, relationships and identities of teachers; it places less emphasis on content and pedagogical content knowledge (Timperley, 2011). Few teachers report experiencing collective participation in traditional PD. Rather, most activities are designed for individual participation (Shields, Marsh & Adelman, 1998). Traditional PD continues to be fragmented, in the format of one-shot workshops at which teachers listen passively to experts speak about topics that are not essential to teaching (National Foundation for the Improvement of Education, 1996). Many traditional PDs do not build on each other; they therefore lack coherence, making it difficult for teachers who have different PD experiences to develop a sense of consistency. For example, in what is widely regarded as one of the best national PD programs implemented in the 1990s, the majority of teacher participants in the Eisenhower Professional Development Program (Garet, Birman, Porter, Desimone & Herman, 1999) reported a lack of coherence in the workshop activities.

Though for the last two decades researchers and policy makers have challenged those offering PD to incorporate the aforementioned elements known to be effective, few programs have done so. (Corcoran, 1995; Darling-Hammond, 1995; Hiebert, 1999; Lieberman, 1996; Little, 1993; Sparks & Loucks-Horsley, 1989). If studies have shown effective strategies, why are these desirable strategies so rare? First, providing activities with multiple high-quality features is challenging (Garet et al., 2001). Providing sustained, intensive PD for all teachers in a school may require more resources than are available, as a substantial amount of lead time and planning is typical for designing and implementing effective PD (Birman et al., 2000; Timperley, 2011). Second, developing and carrying out effective PD is expensive (Jacob & McGovern, 2015). It is estimated, according to Birman et al. and Garet et al. (2001) that effective professional development strategies usually cost more than twice what many school districts can typically afford (Birman et al., 2000; Garet et al., 2001). Third, merely describing effective practices is not sufficient for leaders and teachers to know what to do (Hatano & Oura, 2003). Fourth, teachers are often busy and loaded with heavy teaching and/ or service responsibilities, hence having little time for PD (Henderson & Dancy, 2007). Fifth, lack of supportive environments, such as unfriendly departmental norms, student resistance, and traditional classroom layout, can all hinder teachers' efforts to implement effective instructional strategies (Henderson, Beach, & Finkelstein, 2011; Henderson & Dancy, 2007). Sixth, given the essential nature of building PD around content, it is generally not practical for a school, or in most cases, even an entire district, to offer a highly effective program, since the critical mass of teachers in subjects other than perhaps language arts or mathematics cannot be met.

MODELING INSTRUCTION

In this section, we give a brief history and description of a specific teaching framework, modeling instruction (hereafter referred to as modeling or MI), and its associated PD (Hestenes, 1987, 1997). Due to key features of its development and approach, PD built around MI naturally possesses many of the effective practices discussed above. As summarized below, MI has a record of success, especially in developing student conceptual understanding, as measured by research-based instruments like the Force Concept Inventory (FCI) (Hestenes, 2000; Hestenes, Wells, & Swackhamer, 1992; Jackson, Dukerich, & Hestenes, 2008). Teaching with MI requires some teacher skills not needed in traditional lecture-based instruction. The mode for acquiring these skills has typically been workshops, with leaders cognizant of effective PD practices.

MI originated in the 1980s through the collaboration of a high school physics teacher and a physics education researcher. The story is told elsewhere (e.g., Wells, Hestenes, & Swackhammer, 1995), but as details of it are particularly pertinent to this chapter, we present some history here. The content was explicitly organized around core scientific models to develop a more coherent understanding of how concepts fit together. Students became directly involved in collecting and analyzing the data on which scientific principles were based. This was an early use of microcomputer-based labs, as they were called at the time (Thornton & Sokoloff, 1990). It was also the origin of student-held whiteboards, now ubiquitous in STEM education. The boards served as focal points for scientific discourse after laboratory explorations and during student problem-solving sessions (Wells et al., 1995).

After a few years of implementation, student conceptual and problem solving gains were significant, and the level of student scientific discourse was impressive (Hestenes, 2000). Through PD workshops, the approach has spread nationwide and even internationally. What began in one teacher's classroom has become the basis for dozens of workshops offered annually in physics, chemistry, and biology. Participants have represented a wide range of backgrounds. Some have held advanced degrees in their content areas, but many have been cross-over teachers who found themselves teaching outside their comfort zones. Some have had decades of classroom experience, while others have only taught a few years. They have come from institutions ranging from small private boarding schools to urban high-needs schools.

The modeling approach for physics instruction, in particular, has been described extensively (Barlow, Frick, Barker, & Phelps, 2014; Brewe, 2008; Hestenes, 1987; Hestenes, 1997; Jackson, Dukerich, & Hestenes, 2008; Wells et al., 1995). Regardless of the content area, the pillar of the pedagogy is the modeling cycle (Hestenes, based upon Atkin & Karplus, 1962). Each cycle consists of two main phases: development and deployment. Development starts with a laboratory investigation or demonstration, followed by small group collaborations, sharing of findings with the whole class, and analysis to reach consensus regarding an overarching model. During deployment, students apply their understanding to new problems and situations.

Underlying this sequence is the idea that experts tacitly use a model-centered strategy for problem solving, and that such a framework will also assist novices. Instead of a formula-centered problem-solving strategy, in which a student searches for variables to enter into an equation chosen from a list, MI seeks to develop expert-like problem solving skills where one identifies the deep structure determining a solution approach (Chi, Feltovich, & Glaser, 1981; Hestenes, 1987). By engaging students in these processes repeatedly, they not only learn scientific content, but they also learn how science is done (Megowan-Romanowicz, 2016). Scientific practices are a key component of the three-dimensional learning emphasized by the Next Generation Science Standards (NGSS Lead States, 2013), making MI a natural fit for addressing them.

Do the expert-like problem solving skills at the heart of MI work for students? For improving student conceptual understanding, the answer is a resounding yes. The FCI, mentioned previously, has long been one of the benchmark instruments for assessing the impact of curricular reforms in physics. Over 20 years of research show that after one year of education from a novice MI-trained instructor, students have significant FCI gains (~20–25%), while students of expert modelers may see gains twice as large (Barlow et al., 2014). This agrees with the finding that the largest normalized gains on the FCI result from non-traditional teaching methods (see Hestenes, 1997; Hake, 1998, for discussion). The effectiveness of traditional lecture-demonstration physics instruction has been found to be largely independent of the instructor's knowledge, experience, or teaching style, whereas student gains in courses utilizing non-traditional approaches may be more dependent on the instructor (Hestenes, 2000). Beyond improved conceptual understanding, other positive findings associated with MI include attitudinal results on instruments like the Colorado Learning Attitudes about Science Survey (CLASS) and the Sources of Self-Efficacy in Science Courses Survey-Physics (SOSESC-P) when used for different populations, including high school and university students (Brewe, Kramer, & O'Brien, 2009; Brewe, Sawtelle, Kramer, O'Brien, Rodriguez, & Pamelá, 2010; Sawtelle, Brewe, & Kramer, 2010).

Developing student skills in making and using models is the central teacher objective in MI. Students construct and employ models to make sense of phenomena and communicate their understanding with each other. The teacher plans these activities, then facilitates and manages their implementation. These activities differ from traditional instruction in many respects. A typical modeling cycle is at least two weeks long, with roughly a week each for model development and model deployment. Additionally, the allocation of time is often quite different in MI than it is traditionally; having students create and apply a model is very different than direct instruction on the same topic. Collaborative student-led classroom discourse is another distinct difference; the teacher's role is facilitator and co-investigator, rather than dispenser of knowledge, as she moderates classroom discourse to have students correctly employ scientific terminology and formulate and evaluate scientific claims. All of this leads to the conclusion that "the most *critical element* in successful implementation of the modeling method is the *skill of teacher in managing classroom discourse*" (Hestenes, 1997, p. 19; italics in original).

It bears repeating that a teacher in an MI classroom has skills not typically emphasized in a traditional classroom. In some respects, this may be a minor change. An instructor, for example, may already dedicate a significant portion of his class to laboratory investigations, and these laboratory activities can be modified so they include greater student decision-making. In other ways, however, a much more significant change is necessary. This is particularly true in terms of guiding the classroom discourse, where a teacher's Socratic questioning skills are a far, far greater component of MI than in traditional instruction.

Modeling Workshops

Historically, the training of teachers in MI has been through PD. The workshop is organized around a set of MI materials proven effective in classrooms, but it is important to emphasize that MI is not a set curriculum, but rather a curricular design framework. Workshop participants alternate between student mode and teacher mode. While in student mode, participants work through modeling activities as would their students. They are instructed to think about and portray common student difficulties with the materials, concepts, and processes. This structure has several benefits, including: (a) thinking deeply about how students process the material, (b) observing unanticipated student difficulties (courtesy of their fellow participants), (c) observing how an experienced modeling teacher guides classroom discourse, (d) experiencing personally some of the frustration students can feel when the rules of school are different than they may have experienced in the past, and (e) experiencing a leveling of the field in the workshop experience. Student mode eliminates the opportunity for overly confident teachers to show off their knowledge and at the same time allows a safe space for less certain teachers to voice their beliefs; if an incorrect idea is voiced in student mode, no one other than the individual saying it knows whether this is a genuinely held (mis) understanding or whether it is an accurate portrayal of a student difficulty. During student mode, teacher-style conversations are banned or at least strongly discouraged, as the interjection of teacher concerns and questions disrupts the flow participants need to observe and experience. An effective practice is to write thoughts for teacher mode on sticky notes and place them to the side for later consideration.

In teacher mode, participants debrief their recent experiences, voice their concerns, and ask questions about instructional choices made by the workshop leaders. The leaders also pose questions to challenge participants to think about the philosophy behind instructional decisions and about how they might adapt the activities for their personal circumstances, such as different student populations, levels of parental involvement, administrative styles, course levels, course goals, scheduling, classroom setup, or available equipment. These discussions serve several purposes, including: (a) assisting participants in clearly seeing the goals the leader had for a particular activity and realizing how those goals informed their instructional choices, (b) alleviating participants' concerns about perceived blocks to their implementation of modeling, (c) emphasizing the adaptability of the modeling approach to varied classroom environments, and (d) sharing technical behind-the-scenes details about equipment or other aspects of the activity crucial to success but not appropriate to discuss in student mode.

The training of teachers to utilize MI is now a national endeavor, with in recent summers more than 60 multi-week PD workshops offered and attended by more than a collective 1,200 teachers (American Modeling Teachers Association

[AMTA], 2018). Just as MI is an evidence-based approach for science instruction, the accompanying workshops are evidence-based in their incorporation of best practices identified for teacher PD. It has been noted by Barlow et al. (2014) that MI's focus on content and teacher engagement in learner-centered pedagogies aligns with key characteristics of effective science teacher PD. The fidelity of MI implementation relies on effective PD, so it is unsurprising that its structure has incorporated such characteristics. The successful dissemination of MI through PD (Lee, Dancy, Henderson, & Brewe, 2012) has been described as a triumph, leading Barlow and co-workers to investigate its impact on teachers' instructional practices. They found that MI professional development succeeds in integrating content learning, active learning, participation, and duration (Barlow et al.). The scale of their investigation was limited, however, to just nine teachers, so they were unable to look deeply at other characteristics of successful PD, such as cohesion, as it pertains to modeling instruction.

Although an extensive body of research investigating PD exists and many MI workshops have been offered and evaluated, what is largely lacking is the analysis of Modeling Instruction vis-à-vis professional development. This is a shocking omission given both the success of the program and the central role that PD plays in preparing teachers to implement MI. What takes place at a MI workshop? What leads to science teachers accepting, learning, and then implementing MI in their classrooms? Are there particular affordances or constraints of MI enabling or hindering its adoption and successful execution? The few investigations on this topic (Barlow et al., 2014; Jackson et al., 2008) have been limited, so there is a need to examine more closely what occurs at an MI workshop. This analysis can not only systemize current exemplary practices, but may also serve to support the design and implementation of other PD efforts. Our work provides an opportunity to reflect on already successful MI professional development practices and consider why they may lead to success in ways traditional PD does not.

LONGITUDINAL STUDY OF PROFESSIONAL DEVELOPMENT AT A MODELING INSTRUCTION WORKSHOP

Context

This investigation examines a three-week physics MI workshop that has trained hundreds of science teachers spanning its first 12 years of operation in central Ohio. Details of the workshop's origins are in Cervenec and Harper (2006). What began as a physics workshop was soon joined by simultaneous chemistry and curriculum development workshops; biology and a supplemental engineering offering were introduced more recently. Over this period, the physics workshop has been offered consistently, but it should be noted that the manner in which it has been implemented has been influenced by the other courses.

In mapping the terrain of PD, Borko (2004) noted activities have been studied on different scales, and she grouped the programs of research into different phases. Phase 1 researchers focus on a single site, typically studying the PD program, teachers as learners, and the relationships between these two. Phase 2 researchers study a single PD program enacted by multiple facilitators at multiple sites exploring the relationships among facilitators, the PD program, and teachers as learners. Phase 3 researchers broaden to compare multiple PD programs enacted at multiple sites, providing the opportunity to also study the broader context (Borko). This study is most closely aligned with Borko's Phase 1 investigation, taking place at a single site. However, this study is also longitudinal, with examination of the program at three different stages: The initial offering (year 1), after the workshop was expanded (year 3), and recent offerings (years 9, 10, and 11). It is also longitudinal in that teacher perspectives are investigated at the start of, during, and after a workshop, and then after their school year. These characteristics allow consideration of those factors present in Phase 2 and Phase 3 studies, which include the role of facilitators and the broader context for the programs.

Methods

Teacher PD takes different forms and may have multiple aims, so methods to investigate the effectiveness of PD must reflect this diversity. Learning outcomes are multi-faceted, and content learning is only part of the whole. While the FCI was administered to teachers before and after the workshop, our primary focus is on the participants' views and attitudes toward the MI experiences, as well as the alignment of such with the goals of the PD. Teachers' voices are therefore prominent and highly valued in our description of workshop experiences. As described above, participant feedback was gathered multiple times, each year of the workshop, using both quantitative surveys and open-response questionnaires, and these findings have been reported on a yearly basis by the program evaluator (Upton, 2014, 2017; Upton & Maack, 2005, 2008; Upton & Ross, 2015, 2016, 2017). In addition, two of the authors (K.H. and T.C) each amassed more than 1500 hours as workshop co-leaders and observers at this workshop. In educational research, practitioners and researchers are often placed in separate categories (Nichols & Cormack, 2017). This chapter bridges the categories, as these authors are situated as practitioners researching their own context and practices for a professional development workshop in which they influenced teachers' classroom methods. Specifically, the examination of these data is done from a phenomenological interpretivist perspective, as the presentation and conclusions drawn are a product of the environment that brought the participating teachers and authors together (Denzin & Lincoln, 2013). The perspectives shared in this contribution are representative of, and prevalent among, all workshop cohorts across multiple years and are grounded in the nature of the workshop.

Participants

As shown in Table 3.1, the initial offering of the workshop was physics MI, with 22 participants. The teacher participants in this first offering, and in subsequent ones, have largely been what Borko describes as "motivated volunteers" (p. 5) seeking to learn and try out new ideas. Throughout the history of the program, a typical teacher participant is best described as white, highly educated (most with a Master's degree or higher), and teaching high school courses in STEM. The gender distribution has been closely split between male and female.

A noteworthy feature of the initial cohort of physics teachers was their considerable experience in the classroom. About half of the participants had 11 or more years of teaching experience, with a quarter of the participants having more than 20 years of experience. Over time, while the workshop participants have remained highly educated, with about two thirds having earned a Master's degree or beyond, they have less classroom experience compared to the early years of the workshop. In year 11, for example, no participant reported more than 10 years of classroom experience, whereas in years 1 and 3 about one-half of the participants reported 10 or more years.

Another defining, and perhaps underappreciated, characteristic of these teachers has been their versatility with teaching different science courses. In considering their teaching loads, it became clear that few were teaching only physics. After the first workshop it was decided to gauge what courses the participants taught the subsequent school year. This documentation found about 30–50% were primarily physics teachers, with a teaching load of three or more physics classes.

TABLE 3.1. Description of Workshop Participants

| | | Period 1 | Period 2 | | Period 3 | |
		Year 1	Year 3	Year 9	Year 10	Year 11
Workshop	Physics	N=22	N=16	N=24	N=18	N=16
	Curriculum	n/a	N=11	N=7	N=8	N=13
	Chemistry	n/a	N=18	N=23	N=16	N=13
	Engineering	n/a	n/a	n/a	n/a	N=11
Education	Masters+	78%	81%	62%	60%	80%
Grade level taught	9–12	95%	98%	93%	83%	93%
Gender	Female	55%	55%	52%	64%	55%
Race, ethnicity	White	77%	94%	100%	90%	98%
	Hispanic	14%	3%	—	—	—
	African-American	9%	3%	—	7%	—
	Asian/Pacific Islander	—	—	—	3%	2%

Many of the participants did not teach any physics (about 20%), and the percentage assigned physical science or chemistry was quite high (above 50%). These characteristics had implications for the teaching and learning in the workshop, the perspective of the participants, and the need to develop transferrable skills, as discussed below.

Workshop Activities

Shifting the focus of consideration from who took this workshop to what they did at this workshop involves examination of the logistical and pedagogical characteristics of the workshop in light of the practices for effective PD as outlined previously: duration, content, active learning, participation, type of activity, and coherence, and how participants have experienced and described these practices at this modeling instruction workshop through the years.

Over the three weeks, we have observed participants undergo significant transformations in their perspective of teaching and learning and a subsequent transformation in their own practices. The idea that participants choose to dive in whole-heartedly to adopt MI and move away from the traditional classroom pedagogy they may have used for years is hard to believe. Yet, after the workshop, nearly all the participants indeed make such a change.

The manner in which this transformation occurs is undoubtedly complex, and participants describe the process using terms like "eye-opening" and "a rollercoaster of emotions" (Upton, 2017, p. 20). As leaders we have observed most teachers move through and are positively affected by the workshop in a similar way. This transformation may be attributed, at least in part, to the common goals and objectives of the participants. The goal-driven model of teacher cognition proposed by Hutner and Markmann (2017) contributes to our understanding of this phenomenon. In this model, a science teacher's instructional practices are an attempt to satisfy one or more of the teacher's goals, and these goals are ultimately the mental constructs leading to action. We are not claiming everyone enters the workshop with identical goals, experiences the workshop in an identical way, or finishes the workshop with the same goals. We are suggesting, however, that many participants with similar goals seem to respond to MI activities in predictable ways that may be attributed to characteristics noted for effective PD. As leaders, as we have become aware of these teachers' perspectives over the years, we have reflected on the workshop activities, their sequence, and their connections to the participants' goals.

Table 3.2 provides a timeline for a typical workshop, along with the perspectives, goals, and concerns commonly held by the participants, as ascertained by common participant conversations and questions, at each stage in the workshop.

At the start of the workshop, participants are curious about a range of issues: how the PD experience will function, who their fellow participants are, who will teach them, what they will learn, etc. Across the history of the workshop, the participants have tended to arrive confident with their skills as science teachers. On

TABLE 3.2. A Typical Workshop Timeline

Time Period	Physics Content	Teachers' Perspective	Teachers'Goals	Teachers' Concerns
Day 1	N/A	Curiosity	Learn *about* a new pedagogy Improve my *student's* conceptual understanding	• Finding value from PD
Start of Week 1	(Student mode) • Scientific Processes, • Constant Velocity • Motion	Participation and Skepticism	Improve my *own* conceptual understanding Evaluate pros and cons of MI	• How can the MI format possibly work logistically?
End of Week 1	(Student mode) • Constant Acceleration	Acceptance and Interest	Continue to learn by participating in MI	
Start of Week 2	(Transition to instructor) • Newtonian Mechanics	Concern	Gain skills to lead a MI classroom	• MI uses skills I do not have. • MI uses labs in new ways.
Week 3	• Energy Conservation	Ownership	Improve skills for leading a MI classroom	• Will MI be accepted by parents, administrators, and students?

pre-workshop surveys, about 70–90% of participants agree or strongly agree with the statement "I have a good understanding of effective questioning techniques and how to use them in the classroom" and nearly all (90–100%) agree or strongly agree with the statement "I have a good understanding of fundamental core content in my discipline." (Upton, 2014, 2017; Upton & Maack, 2005, 2008; Upton & Ross, 2015, 2016, 2017). However, they are generally seeking to learn new ideas or skills they can bring into their classes, or perhaps only seeking to learn about new ideas or skills without necessarily using them in their classes. After all, they are confident in their abilities as science teachers and are not eager to discard what works with their students. Often, they value conceptual understanding and are seeking new ways to incorporate it into their classes, but they may be doubtful as to whether these goals can be achieved at a PD workshop. Most have attended (many) traditional PD workshops, and anticipate this MI workshop will be similar, i.e., there may be a few ideas they like and choose to adopt, but they are usually not anticipating a dramatic change to their classes. They already teach these topics, and use similar laboratory experiments, so why would things be that different? These perspectives are consistent with a view of teaching as a profession in which teachers are routine experts who learn how to apply a core set of skills and routines with greater fluency and efficiency (Dall'Alba & Sandberg, 2006;

Hatano & Oura, 2003; Timperley, 2011). Skill development for a routine expert takes place incrementally, in a stepwise, cumulative manner, with an emphasis on procedural efficiency. Traditional PD often is aligned with improving routine expertise, seeking to provide teachers with another skill to add to their repertoire, with the notion that teachers only face a lack of knowledge of effective teaching practices (Timperley, 2011). It is predictable that MI workshop participants arrive expecting a similar experience.

Not every teacher arrives with the preconception a modeling instruction workshop will be like other PD experiences. Some certainly do arrive at the workshop with the inkling their experience will be different, and this number has grown over time. By year 11 of the workshop, about 50% of the participants had been encouraged by a former participant to attend (Upton, 2017, p. 3). This is not to say they anticipate modeling instruction skills will be easily acquired. As a teacher with some insights into MI candidly remarked at the start of a workshop:

> This is going to be a huge stretch for me. I have used very traditional teaching methods in the past and I WANT to use more modeling type methods, which is why I am here. I think it is something I will have to become more comfortable with, and I am prepared to be uncomfortable for a while. (Upton, 2017, p. 21)

The workshop content is introduced by placing the participants in student mode. Active learning begins from the start with the integration of content and pedagogy, making the teachers active participants rather than passive observers. The activities are collaborative, allowing space for the teachers to improve their own conceptual understanding of physics. Evidence for this improvement is found in the teacher's pre and post scores on the FCI, with average normalized gains of 0.20 to 0.40 typical for a given cohort after the 3 weeks.

As the first week progresses, learning objectives in a MI classroom are communicated to teachers through their active participation in the class. These may be objectives the teachers value, such as students designing experiments, using proper scientific terminology, constructing arguments, or offering explanations, but have not regularly included in their own courses. On pre-workshop surveys, less than half of the teachers report their students engage in reflective thinking or writing, or write about their reasoning to solve scientific problems, or learn by inquiry. Other teachers perceive that the MI classroom is consistent with what they have sought to achieve in their own classroom, but now they see this may be implemented using a MI storyline. As a teacher commented, "I feel at home... [MI] is MY style before I knew what modeling was" (Upton, 2017). This may lead to greater interest in MI, but also skepticism as the teachers critically evaluate the MI activities. Will such activities work with my students? How can there be enough time in a real classroom to have these conversations? A teacher raised logistical questions like these when asking, "I'd like to hear how teachers deal with the huge amount of content while maintaining a modeling approach. What do they speed up? How do they incorporate questioning while lecturing through topics?" (Upton, 2014, p. 25).

Near the end of the first week, many participants are convinced MI has value and they are interested in incorporating at least some elements of it into their classroom. They have been learning both content and pedagogy and see the workshop's pattern: they will be in student mode, work collaboratively, observe the leaders, then debrief in teacher mode. Throughout this week, the workshop leaders have been demonstrating adaptive expertise, rather than routine expertise. As adaptive experts they are deeply knowledgeable about both the content and how to teach it, and are experts in retrieving and applying their professional knowledge based on the needs participants present to them (Timperley, 2011). The leaders clearly have the ability to identify when known routines are not optimal, and can shift the class in a new direction.

We have repeatedly observed the perspectives of many participants shift to concern sometime in the second week when they learn that they will now take turns co-teaching lessons. Concern is an understandable response. The classroom discourse style is new to many, with leaders using Socratic questioning to guide the class to a particular objective. Even class components with which the participants are familiar, like labs, are being used in different ways. Because the leaders have mastered the methodology, they guide the classroom activities so smoothly that many participants do not initially realize the leaders are utilizing skills they (the participants) have not used before. The participants may be routine experts, but not necessarily adaptive ones. Participants now engage in meaningful planning and practice mimicking an actual classroom and receive direct feedback. This is the ultimate in active learning. A teacher described this stage of the workshop as the most useful part: "the chance to experience effective modeling techniques and questions, and then the chance to practice using those techniques," or as another teacher wrote, "I really felt that student mode will help me to be successful. Also, my chance to question other 'students' with suggestions was very successful" (Upton, 2014, p. 8). This development continues into the third week, when it is again stressed that MI is not a set of lesson plans to follow, but rather a framework to provide participants with adaptive expertise. This is particularly valuable because most workshop participants teach multiple science courses.

During this final portion of the workshop, the goal is for participants to take ownership of their newfound identity as Modelers and address concerns they may have with its implementation at their school. How will MI be viewed by the various stakeholders, including administrators, colleagues, students, and parents? The leaders, who also have struggled with this question and have first-hand experiences to share, are paramount at this stage. Inclusion of scientific education research articles in the workshop also helps convince teachers that MI is feasible in their classes. As a participant noted:

> Bringing in articles and research was a great way of convincing me early on that this is a completely viable and research-backed pedagogy. Now I have talking points

based on more than anecdotes when I'm talking with administrators and other teachers in my building. (Upton, 2017, p. 17).

As another remarked, "I feel more confident asking administration for materials. I know why I need them, where I will use them, and have data to back up requests." (Upton, 2017, p. 20).

After the workshop, participants hold high and improved opinions of themselves as teachers. For every workshop examined here, all participants reported they benefited from interactions with workshop leaders, interactions with other participants, and the modeling of inquiry. In terms of self-reported gains, 99% of the participants held they had improved their content knowledge, 95% gained skills in complex thinking and reasoning, 98% increased their ability to see connections between math and science, 100% increased understanding of inquiry based instruction, and 86% increased understanding of alternate assessments. With implications for transferring their skills into other classes, 100% of the participants held they had a better understanding for teaching their physics classes, 91% for teaching physical science courses, and 50% for teaching chemistry courses.

Beyond these numbers, teachers are usually confident to begin a new school year with modeling instruction. Many teachers contrast MI with their understanding of traditional instruction, and are eager to implement what they perceive as a superior approach. As one participant described it, "I am excited to integrate this model into my curriculum because I have seen the value of it and the ineffectiveness of alternative approaches first hand. I feel confident" (Upton, 2017, p. 20). Another participant remarked, "Experiencing a new method of teaching made me feel I owe my former students an apology. I should have gone through this process a lot sooner in my career" (Upton, 2017, p. 20).

The logistical and pedagogical characteristics of this workshop exemplify the practices identified for effective PD. Common to all three periods has been the workshop duration, i.e. a 3-week framework with an on-site working lunch for teacher-to-teacher and teacher-to-instructor networking. As the workshop has grown, these networking opportunities have also been enriched, serving as a prime of example of how duration affects other workshop features. With a longer experience, workshops are more likely to include other desirable features (Birman et al., 2000). As a teacher noted "it was important to see the whole process of modeling, from procedure lab to white boarding, to worksheets, and finally deployment lab. The modeling method is a complete package and that is what makes it successful" (Upton, 2014, p. 8). Teachers can experience the complete package of modeling because the workshop has a sufficient duration.

Utilization of student mode as an active learning strategy is valued by the participants. As a teacher remarked, "I found the most value in working through the curriculum as students, making mistakes that we believed students would make and seeing how our facilitators would work through those problems" (Upton, 2017, p. 16). As another noted, a traditional lecture may have been less effective in demonstrating how modeling works, and hence less effective in persuading her to use modeling in her own class:

> I loved that the Modeling Workshop was in no way a lecture based, let me-tell- you-what-modeling-is approach. I don't feel I would have taken away so much information, evidence, and encouragement to begin implementing this method in my classroom if we (the participants) did not have to be in student mode throughout the workshop. (Upton, 2017, p. 16)

The duration of this MI workshop allows participant immersion in an active learning environment that blends content and pedagogy. Teachers who are pragmatically motivated to participate in PD and increase their understanding of a wide range of topics are well served by a workshop framed by a storyline of model development, and in which they play the role of students, albeit at a greatly accelerated pace. For some participants, the inclusion of even more content would be desirable, especially if it led to a greater understanding of how to address student difficulties. As a teacher commented, "I'd like to see even more topics. I liked how they would introduce a topic, show us the pitfalls and expected student responses, and how to guide them toward the correct questions to learn the topic" (Upton, 2014, p. 24). Note also that this teacher describes teaching and learning in a particular way, i.e., being aware of typical student preconceptions and having students articulate this understanding (the expected student response), followed by *guiding them toward the correct questions to learn the topic*. Gaining this perspective of teaching and learning, and acquiring the skill and confidence to implement it, is supported by participants spending time in both student and teacher modes. A teacher speaks to both of these roles when describing his view of the workshop's strengths: "being in student mode and navigating the course through what my students will experience. This was further enhanced with constant reflection in teacher mode. Also, the emphasis of having us do questioning exercises was useful." (Upton, 2014, p. 8). Once again, duration is shown to be a characteristic of effective PD as it allows for the inclusion of other effective practices as well.

Participant networking is an almost inevitable result of the duration and logistical framework, which leads to collective participation on several levels: sharing perspectives while learning modeling techniques, discussing common experiences as science teachers, gaining insights from veteran modelers who lead the workshop, and contributing to a shared professional culture. Participants often describe a significant need to communicate with colleagues that share their professional goals, and sometimes indicate interactions at their own school may not meet this need. For example, a participant described teaching as being "insular during the school year" (Upton & Maack, 2008, p. 12). Consider also this feedback:

> The most helpful part of the workshop was interacting with other science teachers dedicated to improving science education in their classes. I work in a very small district, and the other two science teachers are older instructors who use a lecture

format. During the school year, I feel like I'm on an island because they cannot offer any helpful advice...seeing experienced teachers use modeling and being able to ask them questions was invaluable to me. (Upton & Ross, 2016, p. 9)

Foremost concerns for this teacher are the small size of her district and the traditional approaches of her colleagues. The professional community established at the workshop transcended constraints she perceived existing at her school. For many, participation in the MI community is valuable and worth sustaining. As one offered, "Getting together with teachers that model is so important. I would love to have regular get-togethers just to keep in touch. In many ways, it is the interactions with other teachers that are the most profound" (Upton, 2014, p. 10). The desire to get together with *teachers that model* speaks to the coherence within the teachers' experience, as they seek to integrate MI with their other experiences and goals as professionals because these experiences and goals often parallel those of their fellow participants.

The high value teachers place on networking, community building, and coherence is also illustrated when teachers indicate these qualities were insufficiently addressed at the workshop. As one participant noted, "this was a very general workshop and the practices would look very different depending on the type of school (urban, suburban, rural)," and as another remarked, "it would be helpful to have presenters from varied educational environments. Some questions I had about implementing this curriculum in a city school were left unanswered because others don't have the same experience" (Upton, 2016, p. 23). Comments like these are valuable formative feedback for leaders in continuing to build and serve a diverse community of modeling teachers.

Teacher's Perspectives on Classroom Implementation

Do the participants' views of teaching and learning at the end of the workshop persist? What happens when these teachers return to their schools? Do they continue to view MI as a valuable approach worth implementing, or do the realities of attempting it in their own schools shake their confidence and dampen their enthusiasm? The opinions of teachers regarding various aspects of teaching and learning, prior to taking the workshop, and then at the end of their school year, are shown in Table 3.3. A persistent concern voiced by teachers at the end of the workshop is whether MI will be accepted by their principal. Lacking this support, MI will be difficult, if not impossible, to implement. It is encouraging, therefore, to find teachers describe their principals as being supportive of innovative approaches to teaching science.

Implementing MI causes teachers to form different opinions of the student and teacher roles in science classes (Table 3.3). There is a shift, from teachers showing students how to solve problems that will be on tests, to classroom dialogue among students and teachers about employing problem-solving skills to discover solutions. As described by the teachers, this is not a minor change. As one noted,

TABLE 3.3. Teacher Opinions of Teaching and Learning, Pre and End of Course

Statement	Year 3		Year 9		Year 10		Year 11	
	Pre	End	Pre	End	Pre	End	Pre	End
My principal is supportive of innovative approaches to teaching science.*								
	87%	100%	58%	91%	100%	100%	86%	83%
Classroom interactions involve a dialogue among students and teachers.								
	66%	79%	50%	91%	66%	81%	81%	92%
Student role is to apply inquiry and problem solving skills to discover solutions to problems.								
	73%	93%	54%	91%	66%	81%	87%	91%
I generally assess students' progress using alternative methods.								
	7%	29%	16%	57%	33%	31%	50%	59%
Good science teachers show students the correct way to answer questions they will be tested on.								
	38%	7%	42%	29%	22%	12%	37%	16%
Students should never leave science class feeling confused or stuck.								
	25%	0%	17%	0%	39%	12%	0%	8%

*Percent agreeing or strongly agreeing with statement.

"I completely changed my approach to teaching. I'm much more student and inquiry focused. I'm all about letting the kids learn through experience, evidence, and discourse as opposed to read and remember." (Upton, 2014, p. 17). Workshop participants describe their teaching as being "entirely changed" (Upton & Ross, 2016, p. 22) or undergoing a "complete makeover" (Upton, 2014, p. 17).

TABLE 3.4. Reported Weekly activities, Pre and End of Course

Statement: What do students do?	Year 1		Year 3		Year 9		Year 10		Year 11	
	Pre	End	Pre	End	Pre	End	Pre	End	Pre	End
Learn by inquiry.*										
	39%	69%	31%	78%	29%	85%	41%	75%	46%	67%
Write their reasoning about how to solve scientific problems.										
	34%	56%	25%	64%	21%	91%	50%	75%	46%	92%
Engage in reflective thinking/writing.										
	34%	56%	37%	43%	8%	52%	50%	75%	42%	58%
Make conjectures and explore problem solving.										
	56%	56%	37%	86%	33%	86%	55%	62%	60%	75%
Use scientific equipment.										
	48%	75%	75%	85%	66%	85%	77%	62%	73%	92%

*Percent agreeing or strongly agreeing with statement.

After the MI workshop, participants' understanding of the practice of teaching and learning has changed, leading them to make changes in their classrooms. A student-centric pedagogy is apparent that increases emphasis on having students learn by inquiry, explain their reasoning to solve scientific problems, and engage in reflective thinking and writing (Table 3.4). These activities lead to improved student conceptual understanding. Consistent with other investigations of MI classes, students in physics courses led by first-year modeling instructors in Period 3 had an average normalized gain of 0.21 on the FCI (Barlow et al., 2014). Students in classes led by participants who took a subsequent MI workshop achieved an average normalized gain of 0.50 (Upton, 2014, 2017; Upton & Maack, 2005, 2008; Upton & Ross, 2015, 2016, 2017). These results support the previous finding that instructor expertise is a key factor when utilizing nontraditional approaches (Hestenes, 1997).

SIGNIFICANCE, CONTRIBUTION, AND LIMITATIONS

The longitudinal nature of this study provides an opportunity to gain insights into professional development surpassing those previously reported for MI (Jackson et al., 2008; Barlow et al., 2014). Data collected over the lifetime of the workshops describe a program with strong and consistent success. This is not a programmatic search for a formula that works, but rather implementing a framework that repeatedly persuades and trains teachers to implement MI in their classrooms. This longitudinal investigation is less about changes occurring during the workshop's lifetime and more about commonalities in terms of participants, their goals, the workshop components, and the teacher's perspectives vis-à-vis characteristics of the workshop.

The significance of this investigation is twofold. First, it provides additional support for the paradigm that effective PD addresses issues of duration, content, active learning, participation, type of activity, and coherence. Second, new insights have been gleaned as to why these elements contribute to successful PD. These findings are based on the long-term investigation of an MI workshop, and while these insights are certainly informed by specific practices at this workshop, one could envision teachers making comments similar to the ones shared here about other PD that also incorporated the elements identified in the literature.

As described above, elements of effective PD are found throughout this MI workshop. Two elements of central importance to this program's success have been the extensive use of student mode and the expertise and credibility of the workshop leaders. Both of these features have been present throughout the workshop's history, and their alignment with effective PD has been discussed above. Student mode supports content learning, active learning, participation, and engages participants in meaningful learning. Workshop leaders' expertise in leading student-mode activities is required for these experiences to be valuable, and the credibility and past experiences of the leaders contribute to the coherence of the PD.

The workshop leaders, and other workshop elements, communicate that MI is not an acquisition of isolated skills. Consistent with Dall'Alba and Sandberg's (2006) description of effective PD, MI promotes understanding embedded within a dynamic, inter-subjective practice. Dall'Alba's model includes two dimensions: skills progression and embedded understanding of practice in the profession. MI workshops address both dimensions. The way in which teachers understand their practice and how this understanding organizes their knowledge and skills is usually absent in traditional PD, and yet it is fundamentally important. Whether teachers view their profession as one of transmitting knowledge or one of facilitating learning affects their practices. Therefore, the key questions are what are the professional skills for a teacher, and how is skillfulness developed? MI extends beyond traditional PD's focus on knowledge building to also develop an understanding of professional practices, including ways to address complexities, ambiguities, and dynamic change. Such practices lead to Timperley's (2011) adaptive expertise, and although it is not named as such in the MI community, her description resonates with our experiences. MI's emphases on process, classroom discourse, discovery, and model building all align with developing adaptive expertise.

The MI workshop described here highlighted various aspects of the teaching profession and integrated these practices coherently. The acquisition of knowledge and skills is accompanied by a deeper understanding of practices and may also be accompanied by socioemotional changes in interest, values, or identity (Hatano & Oura, 2003). This must be negotiated by teachers, in terms of goals they have both in the classroom and outside of it (Hutner & Markman, 2017). As they reconceive their profession, teachers grapple with a host of issues that manifest themselves in different perspectives, goals, and concerns that arise throughout the workshop. With this longitudinal study, it has been possible to describe how teachers' views change over a three-week MI workshop, how they understand the practices of their profession, and how they articulate their goals as teachers.

The representative participant comments shared above are consistent with our observations that perspectives frequently evolve in similar ways, beginning with curiosity, shifting to participation and skepticism as student mode begins, acceptance and interest as they find value in MI, concern when they shift to teaching MI lessons, and finally ownership as they formulate plans to implement MI. This progression, which is intertwined with the workshop's pedagogy and logistics, has been consistent across the program's history. We are not claiming this sequence is characteristic of all effective PD; in fact, whether such a sequence is common to other MI workshops is an open question. However, given the degree to which we can connect specific transitions to particular workshop elements, we hold any sequence depends on features particular to a given workshop. It is quite possible that non-Modeling PD could include such a sequence if it shares similar design characteristics and a similar population of participants. This is a topic worthy of further investigation. A consensus has emerged regarding characteristics of effective PD. This investigation suggests that a closer look is warranted to see how participants are affected by these elements during the duration of a PD workshop.

This work indicates that duration is a key element in effective PD, and that activities of insufficient time are unlikely to provide teachers with opportunities to engage in deep discussions or try new practices and obtain feedback (Birman et al., 2000; Garet et al., 2001). If adaptive expertise is a desired outcome, this objective has no shortcut. The manner in which teachers' perspectives changed at this MI workshop, while reproducible in some respects, is undoubtedly complex. Our experiences with MI suggest that elements of effective PD, in order to be effective, only become coherent and convincing for the participants if presented over a sufficient duration.

A limitation of this study is the self-selected population of motivated volunteers who enrolled in these workshops. While we are willing to claim duration is a crucial element of effective PD, the notion that participants undergo a similar evolution of views throughout this MI workshop includes the caveat that they have comparable views and similar motivations at the start of the workshop. This limitation does not mitigate the finding that PD workshops based on known best practices effectively transform teachers' views and practices. It provides additional support for the claim that such best practices are not easy to implement. This investigation provides no reasons to suggest MI would be as compelling if it were introduced in a traditional PD format (though one wonders how it would be possible to frame it traditionally). Rather, this study suggests teacher's perspectives shift in complex ways when PD is of sufficient duration and includes elements like content learning, active learning, and coherence. Without these effective practices, it is difficult to conceive how the shifting of teacher views and goals can occur. This work, therefore, reinforces both the consensus of what works in PD and also the consensus for why the status quo does not.

REFERENCES

American Modeling Teachers Association. (2018). *What is AMTA?* Retrieved from http://modelinginstruction.org

Atkin, J. M., & Karplus, R. (1962). Discovery or invention? *The Science Teacher, 29*(5), 45–51.

Ball, L. D. (1996). Teacher learning and the mathematics reforms: What we think we know and what we need to learn. *Phi Delta Kappan, 77*(7), 500–508.

Barlow, A. T., Frick, T. M., Barker, H. L., & Phelps, A. J. (2014). Modeling instruction: The impact of professional development on instructional practices. *Science Educator, 23*(1), 14.

Birman, B. F., Desimone, L., Porter, A. C., & Garet, M. S. (2000). Designing professional development that works. *Educational Leadership, 57*(8), 28–33.

Borko, H. (2004). Professional development and teacher learning: Mapping the terrain. *Educational Researcher, 33*(8), 3–15.

Brewe, E. (2008). Modeling theory applied: Modeling Instruction in introductory physics. *American Journal of Physics, 76*(13), 1155–1160.

Brewe, E., Kramer, L., & O'Brien, G. (2009). Modeling instruction: Positive attitudinal shifts in introductory physics measured with CLASS. *Physical Review Special Topics-Physics Education Research, 5*(1), 1–5

Brewe, E., Sawtelle, V., Kramer, L. H. O'Brien, G. E., Rodriguez, I., & Pamelá, P. (2010). Toward equity through participation in Modeling Instruction in introductory university physics. *Physical Review Special Topics-Physics Education Research, 6*(1), 1–12

Cervenec, J., & Harper, K.A. (2006). Ohio teacher professional development in the physical sciences. *AIP Conference Proceedings, 818*(3), 31–34.

Chi, M., Feltovich, P., & Glaser, R. (1981). Categorization and representation of physics problems by experts and novices. *Cognitive Science, 5*(2), 121–152.

Cobb, P., Wood, T., Yackel, E., Nicholls, J., Wheatley, G., Trigatti, B., & Perlwitz, M. (1991). Assessment of a problem-centered second-grade mathematics project. *Journal for Research in Mathematics Education, 22*(1), 3–29.

Cohen, D. K., & Hill, H. C. (1998). *Instructional policy and classroom performance: The mathematics reform in California (RR-39).* Philadelphia, PA: Consortium for Policy Research in Education.

Corcoran, T. B. (1995). *Transforming professional development for teachers: A guide for state policymakers.* Washington, DC: National Governors Association.

Dall'Alba, G., & Sandberg, J. (2006). Unveiling professional development: A critical review of stage models. *Review of Educational Research, 76*(3), 383–412.

Darling-Hammond, L. (1995). Changing conceptions of teaching and teacher development. *Teacher Education Quarterly, 22*(4), 9–26.

Denzin, N., & Lincoln, Y. S., Eds. (2013). *The landscape of qualitative research.* Thousand Oaks, CA: Sage Publications Ltd.

Fennema, E., Carpenter, T. P., Franke, M. L., Levi, L., Jacobs, V. R., & Empson, S. B. (1996). A longitudinal study of learning to use children's thinking in mathematics instruction. *Journal for Research in Mathematics Education, 27*(4), 403–434.

Garet, M. S., Birman, B. F., Porter, A. C., Desimone, L. M., & Herman, R. (1999). *Designing effective professional development: Lessons from the Eisenhower Program and technical appendices.* Washington, DC: American Institutes for Research.

Garet, M. S., Porter, A. C., Desimone, L., Birman, B. F., & Yoon, K. S. (2001). What makes professional development effective? Results from a national sample of teachers. *American Educational Research Journal, 38*(4), 915–945.

Hargreaves, A., & Fullan, M. G. (1992). *Understanding Teacher Development.* London, UK: Cassell.

Hake, R. R. (1998). Interactive-engagement versus traditional methods: A six-thousand-student survey of mechanics test data for introductory physics courses. *American Journal of Physics, 66*(1), 64–74.

Hatano, G., & Oura, Y. (2003). Commentary: Reconceptualizing school learning using insight from expertise research. *Educational Researcher, 32*(8), 26–29.

Henderson, C., Beach, A., & Finkelstein, N. (2011). Facilitating change in undergraduate STEM instructional practices: An analytic review of the literature. *Journal of Research in Science Teaching, 48*(8), 952–984.

Henderson, C., & Dancy, M. H. (2007). Barriers to the use of research-based instructional strategies: The influence of both individual and situational characteristics. *Physical Review Special Topics-Physics Education Research, 3*(2), 1–14.

Hestenes, D. (1987). Toward a modeling theory of physics instruction. *American Journal of Physics, 55*(5), 440–454.

Hestenes, D. (1997). Modeling methodology for physics teachers. *AIP conference proceedings, 399*(1), 935–958.

Hestenes, D. (2000). *Findings of the modeling workshop project, 1994–2000.* Retrieved from http://modelling.asu.edu/R&E/ModelingWorkshopFindings.pdf

Hestenes, D., Wells, M., & Swackhamer, G. (1992). Force concept inventory. *The Physics Teacher, 30*(3), 141–158.

Hiebert, J. (1999). Relationships between research and the NCTM standards. *Journal for Research in Mathematics Education, 30*(1), 3–19.

Hutner, T. L., & Markman, A.B. (2017). Applying a goal-driven model of science teacher cognition to the resolution of two anomalies in research on the relationship between science teacher education and classroom practice. *Journal of Research in Science Teaching, 54*(6), 713–736.

Jackson, J., Dukerich, L., & Hestenes, D. (2008). Modeling instruction: An effective model for science education. *Science Educator, 17*(1), 10.

Jacob, A., & McGovern, K. (2015). *The mirage: Confronting the hard truth about our quest for teacher development.* Brooklyn, NY: TNTP.

Kennedy, M. M. (1998). *Form and substance in in-service teacher education.* Arlington, VA: National Science Foundation.

Lee, M., Dancy, M., Henderson, C., & Brewe, E. (2012). Successes and constraints in the enactment of a reform. *AIP Conference Proceedings. 1413*(1). 239–242.

Lieberman, A. (1996). Practices that support teacher development: Transforming conceptions of professional learning. In M. W. McLaughlin & I. Oberman (Eds.), *Teacher learning: New policies, new practices* (pp. 185–201). New York, NY: Teachers College Press.

Little, J. W. (1993). Teachers' professional development in a climate of educational reform. *Educational Evaluation and Policy Analysis, 15*(2), 129–151.

Loucks-Horsley, S., Hewson, P. W., Love, N., & Stiles, K. E. (1998). *Designing professional development for teachers of science and mathematics.* Thousand Oaks, CA: Corwin Press.

Megowan -Romanowicz, C. (2016). What is Modeling Instruction? *NSTA Reports, 28*(1), 3.

National Foundation for the Improvement of Education. (1996). *Teachers take charge of their learning: Transforming professional development for student success.* Washington, DC: Author.

Newmann, F. M., & Associates. (1996). *Authentic achievement: Restructuring schools for intellectual quality.* San Francisco, CA: Jossey-Bass.

NGSS Lead States. (2013). *Next generation science standards: For states, by states.* Washington, DC: The National Academies Press.

Nichols, S., & Cormack, P. (2017). *Impactful practitioner inquiry: The ripple effect on classrooms, schools, and teacher professionalism.* New York, NY: Teachers College Press.

Reimers, J. E., Farmer, C. L., & Klein-Gardner, S. S. (2015). An introduction to the standards for preparation and professional development for teachers of engineering. *Journal of Pre-College Engineering Education Research (J-PEER), 5*(1), 5.

Richardson, V. (Ed.). (1994). *Teacher change and the staff development process: A case in reading instruction.* New York, NY: Teachers College Press.

Sawtelle, V., Brewe, E., & Kramer, L.H. (2010). Positive impacts of modeling instruction on self-efficacy. *AIP Conference Proceedings. 1289*(1), 289–292.

Shields, P. M., Marsh, J. A., & Adelman, N. E. (1998). *Evaluation of NSF's Statewide Systemic Initiatives (SSI) Program: The SSIs' impacts on classroom practice.* Menlo Park, CA: SRI.

Sparks, D., & Loucks-Horsley, S. (1989). Five models of staff development for teachers. *Journal of Staff Development, 10*(4), 40–57.

Stiles, K., Loucks-Horsley, S., & Hewson, P. (1996, May). *Principles of effective professional development for mathematics and science education: A synthesis of standards,* NISE Brief (Vol. 1). Madison, WI: National Institutes for Science Education.

Supovitz, A. J., & Turner, M. H. (2000). The effects of professional development on science teaching practices and classroom culture. *Journal of Research in Science Teaching, 37*(9), 963–980.

Thornton, R. K., & Sokoloff, D. R. (1990). Learning motion concepts using real-time microcomputer-based laboratory experiments. *American Journal of Physics, 58*(9), 858–867.

Timperley, H. (2011). *A background paper to inform the development of a national professional development framework for teachers and school leaders.* Melbourne Australia: The Australian Institute for Teaching and School Leadership (AITSL).

Upton, J. (2014). *Modeling instruction for physical science and chemistry in Ohio, 2013–2014, Evaluation Annual Report.* Retrieved from http://modelinginstruction.org/effective/publications.

Upton, J. (2017). *Modeling instruction for physical science and chemistry in Ohio, Evaluation report: Pre-Survey and end-of-workshop survey results.* Retrieved from http://modelinginstruction.org/effective/publications.

Upton, J., & Maack, S. C. (2005). *Central Ohio Physical Science Modeling Workshop in Ohio, 2004–2005, Evaluation report.* Retrieved from http://modelinginstruction.org/effective/publications.

Upton, J., & Maack, S.C. (2008). *Modeling instruction for physical science and chemistry in Ohio, 2007–2008, Evaluation Annual Report.* Retrieved from http://modelinginstruction.org/effective/publications.

Upton, J., & Ross, R. (2015). *Modeling instruction for physical science and chemistry in Ohio, 2014–2015, Evaluation annual report.* http://modelinginstruction.org/effective/publications.

Upton, J. & Ross, R. (2016). *Modeling instruction for physical science and chemistry in Ohio, 2015–2016, Evaluation annual report.* Retrieved from http://modelinginstruction.org/effective/publications.

Upton, J. & Ross, R. (2017). *Modeling instruction for physical science and chemistry in Ohio, 2016–2017, Evaluation annual report*. Retrieved from http://modelinginstruction.org/effective/publications.

van Driel, J. H., Beijaard, D., & Verloop, N. (2001). Professional development and reform in science education: The role of teachers' practical knowledge. *Journal of Research in Science Teaching, 38*(2), 137–158.

Wells, M., Hestenes, D., & Swackhamer, G. (1995). A modeling method for high school physics instruction. *American Journal of Physics, 63*(7), 606–619.

Wood, T., & Sellers, P. (1997). Deepening the analysis: A longitudinal assessment of a problem based mathematics program. *Journal for Research in Mathematics Education, 28*(2), 163–186.

CHAPTER 4

CO-CONSTRUCTING MODELS THROUGH WHOLE CLASS DISCUSSIONS IN HIGH SCHOOL PHYSICS

Grant Williams and John Clement

We identified modeling practices during whole class discussions of electricity concepts in the classes of two exemplary high school teachers. Four major model construction practices were shared between the teacher and the students referring to observations (O), and generating (G), evaluating (E), and modifying (M) explanatory models. Both groups achieved similarly impressive gain differences over a control group, and high rates of student contributions to modeling, indicating it is possible to achieve the latter. The teachers exhibited substantially different frequencies of scaffolding the practices. We conclude teachers may vary in their level of scaffolding but still experience equally strong student participation in modeling and gains in conceptual understanding. Importantly though, both teachers were focused on fostering the four modeling practices. We provide micro-analyses of classroom transcripts and representative diagrams to illustrate their process of teacher-student co-construction.

Physics Teaching and Learning: Challenging the Paradigm, pages 85–109.
Copyright © 2019 by Information Age Publishing

Keywords: model-based teaching, co-construction, classroom discussion, scientific practices, scaffolding strategies, physics learning

The paradigm of physics teachers as disseminators of abstract conceptual content unlocking the mysteries of the universe by passing on their knowledge through lectures and notes is being challenged by current physics education research. In this study, we analyzed the contributions of both teachers and students during the construction of explanatory models for concepts in circuit electricity in the classes of two exemplary high school physics educators. This analysis attempts to document whether students can contribute significantly to such discussions with model construction practices. We also compare two exemplary teachers on the different degrees of scaffolding they employed for these practices and their corresponding levels of student participation and learning gains. We are interested in this type of student-centered, constructivist physics teaching because it challenges the paradigm of traditional didactic, lecture-based knowledge dissemination often associated with physics teaching and learning.

THEORETICAL FRAMEWORK

One of the core scientific and engineering practices identified by the Next Generation Science Standards (NGSS) (NGSS Lead States, 2013) to help learners construct understandings of difficult concepts is the development and use of models. More detail, however, is needed on the nature of modeling practices. The focus of this study is on student modeling practices visible in whole class discussions and the teacher scaffolding strategies supporting them. (Because the modeling practices we are looking at here are all mental processes, we will use the terms modeling practice and modeling process interchangeably.) A number of researchers have advocated whole class discussions as an effective means for facilitating the construction of scientific knowledge and teaching with a focus on discussion can improve students' scientific reasoning ability and foster conceptual change processes (Hogan, Nastasi, & Pressley, 2000; Lehesvuori, Viiri, Rasku-Puttonen, Moate, & Helaakoski, 2013; Windschitl, Thompson, & Braaten, 2008). Such conversational interaction among teachers and students is thought to provide a means for students to collaboratively construct increasingly sophisticated scientific models through cycles of developing, communicating, evaluating and revising them (Schwarz, Reiser, Davis, Kenyon, Achér, & Fortus, 2009).

In the context of this study, a model is considered to be a simplified representation of a system, which concentrates attention on specific aspects of the system (Ingham & Gilbert, 1991). Models are central to an understanding of underlying mechanisms in science. In the study of physics, in particular, models of various types (physical models, diagrams, equations, graphs, simulations, etc.) can be helpful for supporting students' understanding of abstract concepts (Brewe, 2008; Hestenes, 1987; Wells, Hestenes, & Swackhammer, 1995). These include concepts such as planetary motion, magnetic fields, and electric circuits. We focus on explanatory models, which can be described as mental representations of often hidden causal or functional mechanisms that can explain why phenomena in a system occur (Clement, 1989; Williams & Clement, 2015).

Most traditional electric circuit instruction emphasizes the application of the Ohm's Law equation I=V/R for the quantitative solution of circuit problems. By contrast, model-based learning approaches can be useful for fostering students' development of deeper conceptual understandings of the relationships between current, voltage and resistance within circuits (Borges & Gilbert, 1999; Dupin & Joshua, 1989; Steinberg & Wainwright, 1993). A modeling approach is used in Steinberg et al.'s (2004) Capacitor Aided System for Teaching and Learning Electricity (CASTLE) curriculum employed by the model-based teachers in this study. The CASTLE curriculum utilizes the introduction of large non-polar capacitors into basic electric circuits as a means for focusing students' attention on the transient states of potential differences existing throughout the circuit. By using the analogy of voltage as a type of pressure existing in the compressible electric fluid of a circuit, students are encouraged to generate explanatory models of dynamic pressure changes occurring throughout the circuit as these capacitors go through their charging and discharging cycles. It is thought this emphasis on the conceptual nature of circuit behavior can be beneficial in addressing the many well-documented misconceptions students bring to the study of circuits (Çepni & Keles, 2006; Korganci, Miron, Dafinei, & Antohe, 2015). The CASTLE curriculum employs the extensive use of other analogies, diagrams and discrepant events to engage students and their teachers in the incremental construction of explanatory mental models for circuit electricity. It is this intended conversational classroom process, the different cognitive levels at which teacher contributions are made, and the actual degrees of contributions made to it by teachers and students in these model-based learning situations that comprise the focus of this study.

STUDY BACKGROUND AND RATIONALE

In an earlier phase of our research (Williams, 2011), we examined an experimental group of approximately 270 high school physics students who were learning about electric circuits through the model-based CASTLE curriculum. They, along with an equally sized control group who learned through traditional instructional methods, completed a 20 question, conceptual, non-quantitative pre-test to gauge their understanding of and reasoning about electric circuits. An identical post-test was administered after the period of instruction, which lasted from 6–8 weeks. Both groups had approximately equal distributions of male and female students. A sample question from the test is included in the Appendix section of the chapter.

The purpose of this paper is not to evaluate the curriculum (see Steinberg, 2008). Nevertheless, we need to briefly present results from this testing in order to describe our reasons for choosing two exemplary teachers for an in depth study of their methods. Understanding more about their scaffolding and discussion leading strategies, and the nature of the student-teacher co-construction process in

TABLE 4.1. Pre and Post Test Conceptual Understanding Scores by Treatment Group

	Control Group n = 262		Experimental Group n = 282	
	Raw Score	Percentage	Raw Score	Percentage
Mean Pre Test Problem Solving Scores	6.59 / 20	32.9%	6.70 / 20	33.5%
Mean Post Test Problem Solving Scores	7.75 / 20	38.8%	11.61 / 20	58.1%
Mean Normalized Problem Solving Score Gains	1.17 / 13.41	8.8%	4.91 / 13.30	36.9%

their classrooms, is the main purpose of the present study. A repeated measures analysis of variance (ANOVA) with an alpha value of 0.05 determined that the students in the model-based learning group experienced significantly greater normalized gains (36.9%) in their levels of conceptual understanding over the course of instruction than their traditionally instructed counterparts (8.8%) as displayed in Table1. Calculations of Cohen's (1992) d indicate the effect size of the experimental treatment (model-based instruction of electricity concepts) on students' circuit problem solving outcomes is 1.293; a relatively large effect based on Cohen's scale.

Traditional approaches expect students to learn by listening and absorbing content. The CASTLE curriculum intends for teachers to hold discussions where students can engage in scientific learning processes (practices) involved in constructing models. We selected the two experimental teachers with the largest normalized gains (as shown in Table 4.2) for deeper case study analysis in order to study their scaffolding methods. In this sense we refer to them as exemplary. Their gains were notably higher, but not significantly higher at the $p=.05$ level, than the rest of the experimental group's gains. Additional prior research is best described in the context of our data analysis section below. Some research has identified the need for teachers to use cognitive discussion leading strategies and questioning strategies, but there has been insufficient research done on what those strategies are, and whether they can elicit student model construction processes.

TABLE 4.2. Pre and Post Test Conceptual Understanding Scores by Case Study Teachers

	Teacher A		Teacher B	
Mean Pre Test Scores	6.45/20	32.3%	6.73/20	33.7%
Mean Post Test Scores	11.80/20	59.0%	12.13/20	60.7%
Mean Normalized Test Score Gains	5.35/13.55	39.5%	5.40/13.27	40.7%

Our research questions were:

1. Can we document whole class discussions in which high school physics students contribute significantly with model construction practices, in addition to the teacher's contributions?
2. Can teachers in model-based physics classes participate in whole-class discussions by using a larger or smaller number of scaffolding moves and still foster high levels of student participation and understanding?
3. Can we describe qualitative differences and similarities in the discussion-based strategies used by the two teachers? Can they both be considered types of co-construction?

STUDY CONTEXT AND SETTING

The study was conducted over a two-year period during which the two selected teachers taught the model based electricity unit three separate times with different groups of students, each time spending approximately seven weeks. Both teachers utilized the same model based curriculum, basic constructivist teaching philosophy, and general classroom structure. Group A consisted of a teacher and his students at a small private suburban high school in New England. Of the 39 students, 28 were enrolled in one of two ninth grade general science classes and 11 were students in an eleventh grade physics class. Of the predominantly Caucasian students, 19 were male and 20 were female. Group B consisted of a teacher and his 69 students at a large public suburban high school in the Midwest. Each of the 69 students, of which 35 were male and 34 were female, was enrolled in one of three ninth grade physics classes. The group was a mix of Caucasian, Asian, Hispanic, and African American students. As shown in Table 4.2, the pretest scores indicated the two groups were closely matched on their average level of prior conceptual knowledge of basic electric circuits.

Both teachers utilized class formats in which students alternated between working in pairs on assembling and testing circuit experiments, completing readings and responses in their student workbooks, drawing color-coded analogical pressure-based diagrams of the circuits and their functions, and participating in whole-class discussions moderated by the teacher. In this study, we focus on the whole-class discussions.

DATA COLLECTION AND ANALYSIS

We first conducted microanalyses of discussions from each teacher to identify the major scaffolding strategies used by the teachers and the major model construction processes used by the students. This analysis was a challenging and lengthy qualitative research task. First, passages of whole-class conversations during which the teachers and their students appeared to be engaged in the construction of explanatory models of electricity were video recorded and later transcribed. In total, approximately 5.5 hours of whole-class discussion for each teacher were analyzed from three different classes in an attempt to reduce the effects any one group of students might have on the results. Passages were chosen from each group that featured whole-class discussions during which students were forming explanatory models for observations made in immediately preceding circuit experiments.

In an effort to develop viable descriptions of the strategies and processes used, we employed a construct development cycle (Miles & Huberman, 1994) leading to the progressive refinement of hypotheses about individual teaching strategies and modeling processes (Engle, Conant, & Greeno, 2007). This consisted of: a) segmenting the transcript into meaningful teacher and student statements as the primary units of analysis, b) making observations from each segment, c) formulating a hypothesized construct for or classification of the strategy behind the statement, d) returning to the data to look for more confirming or disconfirming observations, e) comparing the classification of the statement to other instances, f) criticizing and modifying or extending the hypothesized category to be consistent with, or differentiated from, other instances, and g) returning to the data again, and so on. Triangulation from multiple indicators in transcripts and from checks on the ability to use the same constructs across problems and subjects were used to improve and support validity.

Initially, this process allowed us to identify a fundamental similarity existing between the instructional methodologies of the two educators. Each teacher appeared to employ strategies of two distinct types; a *Dialogical* type in which strategies are intended to support students' general engagement in scientific conversation, and a *Cognitive Model-Construction* type with strategies intended to foster students' construction of explanatory mental models. Research by van Zee and Minstrell (1997), Hogan and Pressley (1997), and Chin (2007) has primarily identified what we refer to as dialogical strategies teachers use in whole class discussions in order to promote student engagement and communication. These include participating mainly as a facilitator in the discussion, restating or summarizing student statements, choosing to not directly challenge incorrect statements, redirecting questions back to students rather than providing answers, focusing attention on conflicts and differences of opinion, and inviting responses to other students' statements.

We describe such dialogical strategies as not aimed at specific processes of conceptual learning, but rather as intended to support conversational interaction in general, encourage increased student participation in the discussion, and foster a classroom culture promoting and encouraging student input. While these dialogical strategies certainly help to develop necessary foundations for effective whole-class discussions, we will not discuss them further in this paper since our interest in this study is to explore the Cognitive Model Construction level where we attempted to identify a collection of specific cognitively-focused teaching strategies, not just for promoting participation, but for promoting reasoning and conceptual understanding through model construction.

The three researchers cited just above as well as Minstrell and Stimpson (1996) and Hammer (1995), also identified a few cognitive strategies such as the use of analogies, inductions, and discrepant questions. Meanwhile, researchers within our own group (Clement & Rea-Ramirez, 2008; Clement & Steinberg, 2002; Khan, 2003; Williams, 2011) began to focus on teacher and student model co-construction in a variety of science learning environments; describing teacher and student statements as contributing to the construction of model elements through a variety of cognitive processes. We have described these processes as being centered on the fundamental modeling practices of experimental observation (O), model generation (G), model evaluation (E), and model modification (M) (Williams & Clement, 2015). Three of these categories (G, E, M) originally grew out of observations of scientifically trained experts thinking aloud about explanation problems (Clement, 1989, 2008). Using this OGEM process framework, we engaged in the construct development cycle described above to develop the following criteria to code both student and teacher statements during whole-class modeling discussions into four categories.

1. Observations (O): The statement either asks for, or provides, observations made or outcomes noted either in a previous classroom experiment or demonstration, an everyday occurrence, a video, or other source. This may be done for the purpose of bringing the attention or memory of the participants to the phenomenon being discussed, or it may be a request or suggestion for designing or doing a future observation(s). Examples of key phrases that help identify observation strategies: "did you see . . . ," "what did you notice...," "tell us about your observations . . .," "what was detected . . .," "what would we see if. . ."etc.

2. Generation (G): The statement either asks for, or provides, a theory, explanatory model or model element, conception, or model based explanation. This can be done with varying degrees of speaker confidence in the correctness of the statement and can be done in either a declarative or interrogative manner. Examples of key phrases that help identify model Generation strategies: "why do you think that happened. . .," "what do you think is happening . . .", "what explanation can you think of for . . .", "I think that maybe what's going on is . . . ," "I think it does that because. ."etc.

3. Evaluation (E): The statement refers to a theory, explanatory model, conception, or model-based explanation that has previously been or is currently under discussion. The statement either asks for, or provides, an evaluation, judgment, refutation, criticism, support, or endorsement of a particular explanatory model. Examples of phrases that help identify model evaluation strategies: "do you agree with . . . ," "what do you think

of that explanation....," "that makes sense . . .," "I also believe it could be. . .," "but that doesn't explain why . . .," "do you think that is the way it works. . . ," etc.

4.   Modification (M): The statement either asks for, or provides a suggested change, revision, adjustment, or modification to a theory, explanation, or explanatory model under evaluation. This may involve only a minor alteration, variation, or addition or could introduce a substantially revised model with little resemblance to the original. Sometimes the modification statement comes with little verbal evidence that an evaluation process has been underway as students often engage in this process internally. If the statement appears to make little or no reference to the previous model, it is instead considered to be in the generation category. Examples of phrases referring to an explanatory model that help identify model modification strategies: "does anyone see it a different way . . .", "would anyone suggest changing . . .", "maybe if we explained it like this . . . ", "could it be more along the lines of . . .", "how could we fix the model so that it considers ...", etc.

In the present study, statements made by the teachers and students during whole class discussions were first examined to see if they fit into the OGEM process pattern, at a macro level we call model construction strategies. Criteria for those strategies were developed and refined. Then, a larger number of teacher micro strategies were identified at a smaller grain size, such as "Teacher provides an analogy" or "Teacher requests (generated by students) a model element." We call these micro strategies nonformal reasoning strategies. We found that each of these micro strategies could be seen as a sub-strategy for one of the four macro OGEM strategies; for example, the above micro strategies can both be seen as ways of contributing to the macro strategy of generating a model (G). Another way to view this is the macro strategies refer to the goals or objectives of the actions taken by teachers while the micro strategies refer to the specific actions taken.

RESULTS

Diagrammatic Representations of the Modeling Discussions

In an attempt to visually portray the interplay between the micro-level strategies and macro-level OGEM processes, we developed a diagramming notation to represent the construction processes teachers and students engaged in during these classroom discussions. We constructed such diagrams for a representative subset of the discussions. In their simplest form, the diagrams are horizontal versions of the classroom transcript with student statements presented above the teacher statements, and time running from left to right. For this reason, the diagrams tend to be wide, and for presentation here, necessitated being split into two parts: a and b. The horizontal strip across the middle of the diagrams contains short written phrases which describe the evolving explanatory models. These phrases represent our hypotheses for the teacher's conception of what a student's addition to the model was at a given point in the discussion, based on the student's statements. It was assumed the teachers were aiming to foster model construction based on their view of the student's model at that time, and how it differed from the target model.

Arrows pointing from both teacher and student statements toward the explanatory model descriptions in the center strip indicate shared contributions to the changes or additions in the models. At other times, arrows from the model descriptions are directed toward teacher statements, indicating the influence of the current model on the teacher's next query or comment. The very general form of this role for the teacher is described by Hogan, Nastasi, and Pressley (2000, p. 405) as the teacher "holding together the threads of the conversation, weaving students' new statements with prior ones to help them link ideas and maintain a logical consistency." This is a skill both educators in this study displayed in their teaching.

Immediately below the teacher statements is a brief description of the hypothesized teaching moves at the micro level of nonformal reasoning strategies. These include such strategies as: teacher requests observations, teacher provides a model element, teacher requests the running of a thought experiment, and teacher provides concept differentiation. Arrows to these micro level strategy descriptions point upward to illustrate their being driven by one of the four macro processes (observation, generation, evaluation, or modification). In Figure 4.1, for example, the 2nd through 6th teacher statements all serve the goal of having students generate (G) a model. One can differentiate, however, between the micro strategies of requesting initiation of model construction, requesting an analogy, and requesting elaboration of the model. These three different micro strategies appear to be contributing to the macro process of model generation (G). The macro level or model construction process layer portrays the larger time scale goals of the teacher in engaging the students in the process of generating an explanatory model. This instance of the generation (G) macro process points to different types of micro strategies portraying the relation specific micro strategies serve a smaller number of more general and longer-duration macro processes.

Above the students' statements on the diagrams, we analyzed each statement to describe their micro level processes in contributing to the model construction processes at the top. As is the case for the teacher strategies, we attempted to link (via arrows) each of these student contributions to the macro level OGEM processes of the model construction process. In what follows, we give an analysis of a discussion led by each teacher, using these model evolution diagrams.

Episode 1: Teacher A

In the experiment prior to the whole-class discussion in Episode 1, the students in Teacher A's class were using magnifying glasses to examine the filaments of two different types of miniature light bulbs they were using in the CASTLE

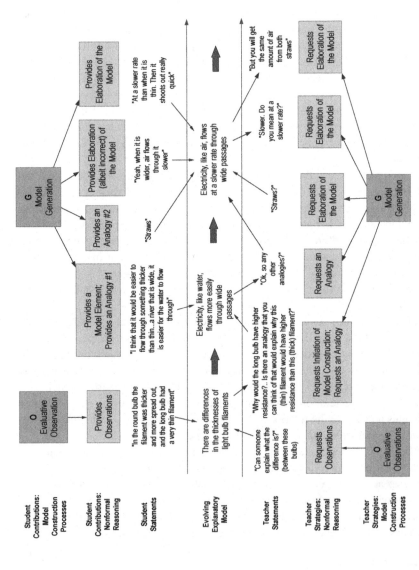

FIGURE 4.1. Whole-Class Model Construction Discussion—Episode 1—Part A

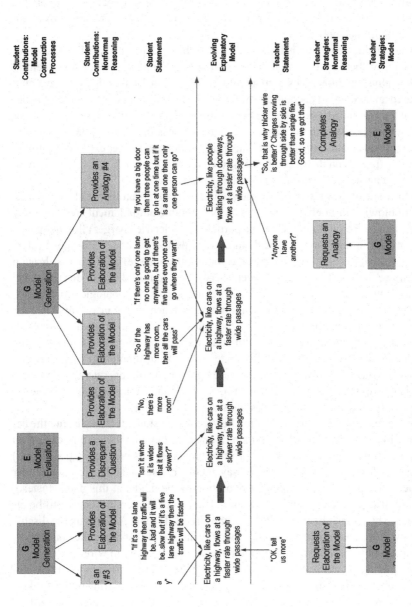

FIGURE 4.2.   Whole-Class Model Construction Discussion—Episode 1—Part B

circuit building kits. By observing the physical differences in the filaments (one thick and the other thin) of these bulbs, it is intended that students will develop explanatory models to account for differences in their behavior and effects on the circuit. The curriculum draws on students' previous experiences to support the development of analogies that can aid in their understanding of charge movement in light bulb filaments.

The episode begins with Teacher A asking the students to share their observations (O) of the light bulb filaments, as shown in Figure 4.1. Once they identify the key physical differences in filament wire thickness, he asks them if they could develop an analogy to account for differences in the resistance to charge flow between the two bulb types. Therein begins the model generation (G) process. After a student responds with the first suggestion of a possible analogy, the teacher encourages additional contributions. This is likely done to further explore the notion of easier flow through wider passages, a concept often confusing the distinction between charge flow rate and charge flow speed.

This issue of flow rate vs. flow speed surfaces through another student explanation of blowing through drinking straws as a suitable bulb filament analogy. Flow rate refers to the total number of air particles (or electric charges) flowing past a certain spot in the straw (filament) in a given period of time. Flow speed refers to the velocity of any one air particle (or electric charge) as it travels through the straw (filament). This is a concept very often confused or not discriminated by students learning physics and one making the use of analogies to describe charge flow in wires challenging without proper teacher guidance. Another issue concerns the initial conditions of the thought experiment in an analogy, such as whether the speeds in different width straws are being compared with the same pressure source, or with the same flow rate source. In an attempt to clarify, Teacher A requests elaboration of the model regarding the issue of flow rate vs. flow speed and later provides a concrete clarification of the model concerning the number of particles (charges) flowing through the straws (filaments).

What results is a rich conversation between three students who dispute the accuracy of the highway analogy. Again, it appears they may be getting caught up on the distinction between flow rate (total number of cars passing by per second) versus the flow speed of each car (in, say, meters per second). The teacher neither requests nor provides any further elaboration of the model at this point. Instead, he asks if there are any other analogies. The analogies here all appear to be attempts to help generate (G) a model for the circuit. Teacher A wraps up the discussion by using the student-suggested doorway analogy to integrate the concepts of passage width and flow rate as applied to charge movement in wires.

Not all ideas put forward are correct. But, what is impressive to us in this episode is the incorporation of student ideas, and the gradual evolution of the models; with evaluations and modifications, yielding competing ideas and subtractions from the model as well as additions, as opposed to a monotonic build-up from instruction. The former process is more like real science than is the latter. While the teacher clearly facilitates this process through the strategic use of scaffolding questions, it is largely the students who are making contributions to this developing model. This is a significantly different process than the traditional teacher-centric approach of information promulgation often occurring in physics classes.

Episode 2: Teacher B

Just prior to the whole class discussion featured in Episode 2, the students in Teacher B's class conducted an investigation in which they first assembled an electric circuit (referred to in the transcript as Circuit A) containing two light bulbs connected in series with a previously discharged 1 Farad non-polar capacitor as shown in Figure 4.3. The purpose of this investigation was for the students to establish that a neutralized or discharged capacitor placed in a circuit without a battery would not result in the lighting of the bulbs. The second part of the investigation involved the insertion of a battery pack into the circuit as shown in Figure 4.4 (Circuit B).

The purpose of inserting the battery pack into a circuit that previously experienced no charge flow was twofold: 1) to cause the discrepant event of the bulbs lighting momentarily and then fading out, and 2) to intentionally support the common misconception that bulb lighting in circuits requires the inclusion of a battery. In the subsequent investigation, the battery pack is removed and the wires reconnected resulting in another discrepant event; the brief re-lighting of the bulbs in a circuit without a battery pack, thus challenging the previous misconception.

After students have investigated both circuits A and B, Teacher B begins the post exploration discussion by having the students reflect on their observations (O) of the circuit building activity. First, he provides an observation by reminding them the light bulbs did come on but, then he quickly turns the discussion over to the students by requesting they provide their own observations, specifically of the duration and brightness of the bulb lighting. When one student reports that the bulb brightness was not constant, Teacher B supports the class's engagement in the model generation (G) process by requesting they provide a model element to explain the behavior of the electric charges in the circuit. After encouraging the

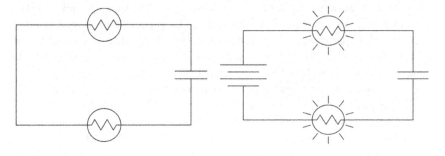

FIGURE 4.3.   Circuit A   FIGURE 4.4.   Circuit B

students to further describe their explanatory models of charge movement, the teacher requests additional observations (O), this time from an earlier exploration. This is likely done for the purpose of making a connection between bulb lighting and compass needle deflection as two types of evidence for charge movement in circuits.

Once the students report their earlier memories of the compass needle deflection, Teacher B refocuses on model generation (G) activity by again requesting the students suggest explanatory models based on their observations. Since the first student response is not as developed as is required, the teacher requests the students evaluate (E) and modify (M) the model under discussion and ultimately add to and improve it, bringing it more in line with the scientifically accepted target model.

At this point in the discussion, Teacher B again focuses his students on the act of model generation (G) by requesting they propose explanatory models based on their observations of Circuit B. Specifically, he guides the students through the logic chain that if a) compass needle deflection occurs when charge is flowing in wires, and that if b) bulb lighting occurs simultaneously with compass needle deflection, then c) when bulbs light, charge must be flowing. Teacher B then supports the students' evaluation (E) of their model by requesting they run a thought experiment predicting what would occur if compasses were used to evaluate the movement of charges in Circuit B where a battery back joined the two light bulbs and capacitor already present. The episode concludes with the generation (G) of a model in which capacitors, in conjunction with a battery, can affect the rate of charge flow in electric circuits. This is an important step toward developing a more generalized explanatory model of differing regions of charge density or electric pressure as causing the movement of charge.

What is most salient about Episode 2 is Teacher B's ability to guide his students in generating explanatory models by developing inferences from their own experimental observations. This activity represents a constructivist approach to learning about charge flow in electric circuits as compared to a more traditional one in which students are first taught the theory then conduct experiments to confirm it. What is also important in this episode are the scaffolding strategies Teacher B utilizes when students' attempts at constructing explanatory models are not as developed or sophisticated as are required to adequately move the process in the direction of the target model. In particular, in part B, the teacher asks the students for experimental evidence to extend the initially proposed explanatory model and secondly, requests refinement of the model by asking for a repair to the language describing the model. These are important strategies because they help the students understand any shortcomings in their own models without directly telling them that they are wrong, serving to encourage them to continue with the model construction process and to see model building is a process of continual evaluation and modification.

We developed the co-construction diagrams above to provide; 1) a visual representation of the interplay between students and teachers in co-constructing explanatory models for scientific phenomena; and 2) a means of interpreting the strategic role of the teacher in scaffolding the observation, generation, evaluation and modification processes of model construction. Vygotsky (1962) referred to the gap between the thinking students can do on their own and what they can do with support from others as the Zone of Proximal Development. Teacher supports helping to bridge this gap are often referred to as scaffolding. In particular, in this chapter, by scaffolding we mean guiding and supporting a discussion with questions, comments, and occasionally ideas contributing to student model construction. Partly because we focused on discussions rather than sections where the teacher gave a presentation, the great majority of the teacher contributions predominantly took the form of questions. The diagrams also 3) portray the relationship between the teacher strategies and student statements at the micro nonformal reasoning level and the macro OGEM model construction processes level.

Student and Teacher Model Construction Participation Ratios

During the 11 hours of whole class discussions facilitated by these two teachers, the students contributed more than 800 instances of these OGEM practices. Students in the transcripts were identified as follows: the first to speak was coded Student 1, and his or her subsequent utterances were attributed to Student 1, the second to speak was coded Student 2, etc. On average, 74% of the students in Teacher A's classes and 66% of the students in Teacher B's classes participated in any one discussion. These values were determined by noting the percentage of students contributing during each discussion session for Teacher A (e.g. 17 out of 24 students present or 70.8%), adding the percentages together, and dividing by the total number of sessions for that Teacher; and similarly for Teacher B.

Students contributed to each of the observation, generation, evaluation, and modification categories. Considerable differences existed between teachers in the raw counts of their conversational statements and in the ratios of the student to teacher contributions. Results for ratios of student and teacher contributions are shown in Tables 4.3 and 4.4. While multiple student or teacher statements at the micro level may contribute to a single OGEM process at the macro level, each of the individual micro-level contributions was counted separately as either an O, G, E or M in the data collection process.

While the rate of student verbalization was only slightly higher in Teacher B's classes than in Teacher A's classes (434 turns compared to 411 turns in the same approximate time), the considerably higher rate of teacher contributions in Group B caused the comparative ratios of student to teacher contributions to be quite different. For example, as shown in Tables 4.5 and 4.6, for teacher A, students contributed 2.6 times as much to model development in the overall aspects of the OGEM processes as the teacher did whereas for Teacher B, students contributed only 1.2 times as much as their teacher did.

It is particularly interesting to compare the sub-category totals to see that, for generating models, students in Teacher B's classes provided roughly twice the

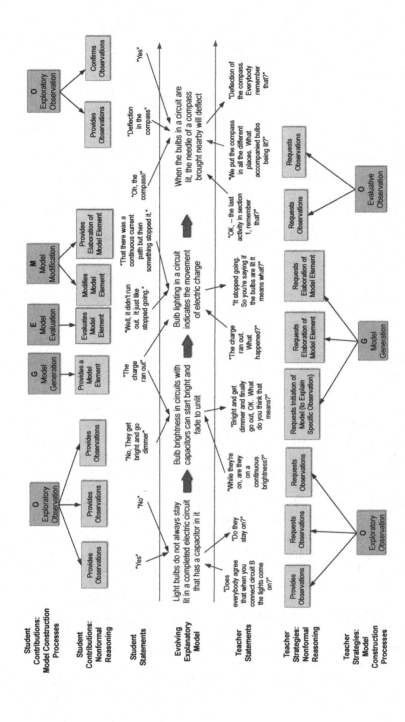

FIGURE 4.5. Whole-Class Model Construction Discussion—Episode 2—Part A

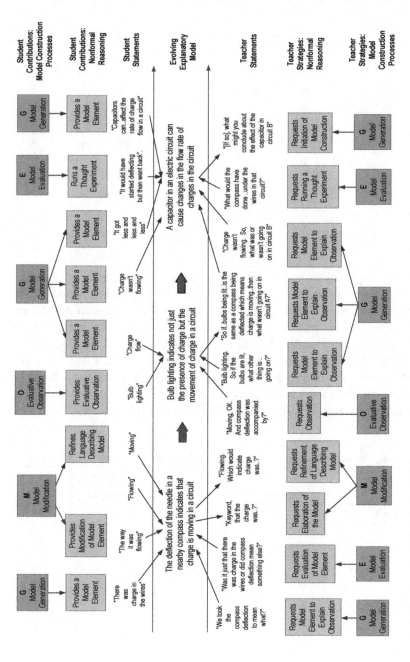

FIGURE 4.6. Whole-Class Model Construction Discussion—Episode 2—Part B

TABLE 4.3. Student and Teacher Contributions to OGEM Processes—Group A

Group A

Discussion #	Students				Teacher			
	O	G	E	M	O	G	E	M
1	21	69	17	4	5	10	16	2
2	24	18	8	4	2	8	10	1
3	11	21	6	2	4	7	6	0
4	17	27	10	3	6	9	8	2
5	22	38	14	6	7	13	15	3
6	19	33	12	5	7	11	7	2
OGEM Process Totals	114	206	67	24	31	58	62	10
Grand Totals	411				161			

number of contributions as the teacher, while in Teacher A's classes student contributions were considerably higher at 3.6 times what the teacher offered. When it came to evaluating models currently under discussion, in Teacher A's classes the student to teacher participation ratio was 1:1, however, in Teacher B's class, the teacher evaluated models 3 times as frequently as students did. As well, in Teacher A's class, students contributed to model modification 2.4 times what the teacher did while in Teacher B's class that same ratio was considerably lower at 1:1. For each of the four OGEM practices, it appears Teacher B chose to play a more active and engaged role in leading the co-construction process.

TABLE 4.4. Student and Teacher Contributions to OGEM Processes—Group B

Group B

Discussion #	Students				Teacher			
	O	G	E	M	O	G	E	M
1	5	50	9	4	10	26	35	2
2	15	61	3	2	7	18	23	2
3	17	49	7	3	11	23	18	4
4	12	33	8	5	9	27	22	4
5	20	54	11	2	8	29	20	3
6	14	37	9	4	6	21	28	4
OGEM Process Totals	83	284	47	20	51	144	146	19
Grand Totals	434				360			

TABLE 4.5.   Student and Teacher Model Construction Contribution Ratios—Group A

	Raw Ratio	Simple Ratio
Observation	114 : 31	3.7 : 1
Generation	206 : 58	3.6 : 1
Evaluation	25 : 26	1 : 1
Modification	24 : 10	2.4 : 1
Overall	411 : 161	2.6 : 1

TABLE 4.6.   Student and Teacher Model Construction Contribution Ratios—Group B

	Raw Ratio	Simple Ratio
Observation	83 : 51	1.6 : 1
Generation	284 : 144	2 : 1
Evaluation	47 : 146	0.3 : 1
Modification	20 : 19	1 : 1
Overall	434 : 360	1.2 : 1

DISCUSSION

Discussion of research question 1: *Can we document whole class discussions in which high school physics students contribute significantly with model construction practices, in addition to the teacher's contributions?* with over 400 student contributions to model construction practices in a total of approximately 5.5 hours of recorded discussions for each teacher, it can be concluded students made a significant contribution to constructing the explanatory models for circuit electricity in these classes. While there were different degrees of scaffolding by the teachers, we are counting instances of relatively high order cognitive contributions on the part of the students, not just recitals of facts. For both teachers, the number of student contributions was greater than the number of teacher contributions. So, there is a body of evidence here that students contributed a significant number of ideas to the model construction process representing a sharp departure from the traditional paradigm of the teacher as the provider of knowledge and the student as the mere recipient. This result resonates with an early study by Hake (1998), in which he identified interactive-engagement approaches to instruction as a key feature successful innovative programs in physics instruction had in common, showing they outperformed others on standardized tests of conceptual understanding. In the present case, however, we also have evidence indicating that students were engaged in, and gaining practice in, the OGEM scientific thinking processes.

Regarding our second research question: *Can teachers in model-based physics classes participate in whole-class discussions by using a larger or smaller number of scaffolding moves and still foster high levels of student participation and understanding?* we also asked what the kinds and quantity of teacher scaffolding moves were for each of these four practices. The strategies identified in the bottom two rows of Figures 4.1, 4.2, 4.5, and 4.6 can be thought of as illustrating a more fine-grained view of scaffolding strategies in model-based discussion leading. For a more complete discussion of these strategies at both macro and micro levels, see Williams and Clement (2015).

If we take the tallied OGEM teacher contributions as instances of scaffolding, it is apparent in Tables 4.3 and 4.4 that Teacher B provided considerably more scaffolding than Teacher A, and this occurred as well within each of the four OGEM categories. The data support our initial impressions from the video recordings of Teacher B's very active involvement in the discussion compared to Teacher A's more reserved style. Although the two teachers' scaffolding styles (as indicated by their frequency of scaffolding OGEM practices) were quite different, they still led to high degrees of participation and learning in their students. This finding suggests there is not one best way to support students' effective engagement in constructing explanatory models for physics concepts.

In response to our third research question: *Can we describe qualitative differences and similarities in the discussion-based strategies used by the two teachers? Can they both be considered types of co-construction?* In all categories of the OGEM modeling process, Teacher A seemed to participate substantially less than Teacher B in doing so, having less than half the overall contributions (161) as compared to those of Teacher B (360) in equivalent time periods. When these are compared to the 400 plus student contributions for each group, we can infer Teacher B's discussions involved more alternating Teacher and Student contributions (TSTS...), whereas Teacher A's involved more Student-Student contributions (TSSSTSS...). Keeley (2008) suggests the analogies of ping-pong or volleyball to describe student and teacher discourse interactions involving TSTS...and TSSSTSS... exchanges respectively. This difference is aptly illustrated in Figures 4.1 and 4.2 vs. Figures 4.5 and 4.6. McNeill and Pimentel (2010) have studied both the advantages of leading discussions with SSS exchanges, and the difficulty of teaching this practice to teachers.

In thinking about hypotheses to explain these differences, the above data resonated with some less formal observations we had made from the video recordings. We noticed Teacher A generally had longer wait time for students to answer after he asked a question, and would sometimes allow SSS exchanges for 30–50 seconds. Another factor discussed by others is the need for both divergent and convergent periods in discussions (Scott, Mortimer, & Aguiar, 2006). While both teachers solicited student ideas in a divergent way, Teacher A sometimes appeared to let this go on for longer periods, as illustrated in Figures 4.1 and 4.2. We also had the impression Teacher B not only used more questions but also used narrower, more specific questions to guide student thinking. As a metaphor, one might describe the teachers' questioning strategies as rungs on a ladder to be climbed, where A's rungs were further apart than B's rungs. For these reasons it is possible students in Teacher A's class were able to practice and learn more thinking skills—in the form of spontaneous model construction (OGEM) processes enacted with less teacher guidance—even though they evidenced the same content gains. We did not, however, have the resources to measure gains in thinking skills directly in this study, so that is an interesting question for future research.

There were some strong similarities between the two teachers. Each fostered equal or greater levels of student participation than that of themselves in most aspects of model co-construction and appeared focused on guiding the discussion enough to converge on target models. From a broader perspective, the two teachers were very close to one another on a spectrum of teaching approaches ranging from didactic and teacher-driven at one end to constructivist and student-centered at the other. As illustrated in Figures 4.1 and 4.2 and 4.5 and 4.6 for both Teacher A and Teacher B, the vast majority of their strategies were in the form of requesting students contribute to the model co-construction process rather than providing such pieces of the puzzle themselves. This is quite different than the discussions occurring in more traditional teacher centered classrooms.

Another shared characteristic of Teacher A's and B's model based instructional efforts is, while both were able to readily engage students in participation in all four of the OGEM processes, the number of student statements in the evaluation (E) and modification (M) categories were relatively low. One explanation for the lower rates of student participation in these aspects of the modeling process is they may have felt the tasks of evaluation (E) and modification (M) belonged in the hands of the teacher, as is typically the case. Another is they may simply had not had much experience with taking the lead on critically evaluating and suggesting revisions to models, posing a significant challenge.

We hypothesize the most important commonality though is both teachers exhibited the qualitative pattern of using OGEM support strategies and, primarily by asking questions, encouraged their students to participate in those same processes. Student model generation, evaluation, and modification processes are not encouraged in traditional instruction so both teachers were unusual in this way. The teachers rather than the students, for the most part maintained control of the direction of the discussion through guiding questions. As the teachers occasionally contributed ideas and terms where needed so, both students and teachers contributed to the discussions. In this way both teachers appeared to be fostering processes of teacher-student co-construction.

CONCLUSION

As we have discussed, results of an initial phase of this research (Williams, 2011) found students in model-based CASTLE classes recorded significantly greater pre- to post-test gains in conceptual electric circuit reasoning and problem solving outcomes than their counterparts who learned the concepts of electricity through more traditional, lecture and equation-based means. As a follow up to these results, we analyzed video recorded episodes from the classes of the two most successful model based teachers in an attempt to identify and describe the types of scaffolding strategies and student modeling practices being employed during large group discussions (Williams & Clement, 2015). There, through microanalysis of protocols, we identified two distinct types of cognitive methods in teachers' repertories; micro non-formal reasoning strategies, and macro model construction processes. At the macro level we identified four major OGEM practices or processes; observation, model generation, model evaluation, and model modification. The intention of the CASTLE curriculum was to foster student involvement in such practices and we asked whether this had in fact happened. We tallied student contributions to these four practices that were contributing to an evolving model.

Students made over 800 contributions of these four model construction practices in 11 hours of discussion, providing more contributions than the teachers in each case which can be a reasonable result as there was only one teacher per classroom. We conclude it is possible to elicit frequent student participation in model construction practices. We also asked what the level of teacher scaffolding was for each of these four practices. While both classes achieved similarly impressive gains, the teachers exhibited substantially different frequencies of scaffolding the practices. While Teacher B chose to be almost equally involved in the co-construction process by contributing as many OGEM moves as his students, Teacher A displayed a much more reticent and reserved style. We conclude teachers may vary in their amount of scaffolding and still experience equally strong student participation in modeling and gains in conceptual understanding. This gives us some insight into the range of teacher-student interaction profiles that can produce exemplary gains. We hope teachers will find these results relevant to decisions about types of scaffolding and the intensity or frequency of scaffolding they provide.

There were important similarities between these successful instructors. Both appeared to be scaffolding the broader processes of observation, model generation, model evaluation, and model modification. Students are typically not encouraged to use the latter three practices in traditional instruction so, we believe these classes were challenging the paradigm of teacher-centric instruction. Since both students and teachers contributed scientific model construction practices to the discussions, we characterize the overall process as one of teacher-student co-construction. This process may be considered a middle ground between purely teacher-generated and purely student-generated models in the classroom. We hypothesize what remained most important was the teachers' ability to foster students' engagement in the four key modeling processes of the activity, in the effort to help students construct meaningful explanatory models for scientific concepts.

This material is based upon work supported by the U.S. National Science Foundation under Grant DRL-1503456. Any opinions, findings, conclusions or recommendations expressed in this work are those of the authors and do not necessarily reflect the views of the National Science Foundation.

APPENDIX: ELECTRIC CIRCUITS PRE/ POST TEST

The following is a sample problem from the pre-post test.

In this circuit, *all four bulbs are identical*, and *all four bulbs are lit*, although they *may or may not all be the same brightness*.

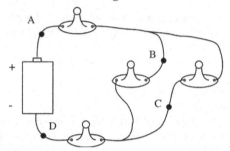

Which of the following is true?

A. The current at point B is greater than the current at point C.
B. The current at point B is equal to the current at point C.
C. The current at point B is less than the current at point C.
D. There is not enough information to know the relative current at the two points

REFERENCES

Borges, A. T., & Gilbert, J. K. (1999). Mental models of electricity. *International Journal of Science Education, 21*(1), 95–117.

Brewe, E. (2008). Modeling theory applied: Modeling instruction in introductory physics. *American Journal of Physics, 76,* 1155–1160.

Çepni, S., & Keles, E. (2006). Turkish students' conceptions about the simple electric circuits. *International Journal of Science and Mathematics Education, 4,* 269–291.

Chin, C. (2007). Teacher questioning in science classrooms: Approaches that stimulate productive thinking. *Journal of Research in Science Teaching, 44*(6), 815–843.

Clement, J. (1989). Learning via model construction and criticism: Protocol evidence on sources of creativity in science. In Glover, J., Ronning, R., & Reynolds, C. (Eds.), *Handbook of creativity: Assessment, theory and research* (pp. 341–381). New York, NY: Plenum.

Clement, J. (2008). *Creative model construction in scientists and students: The role of imagery, analogy, and mental simulation.* Dordrecht: Springer. Softcover edition, 2009.

Clement, J., & Rea-Ramirez, M. A. (Eds.). (2008). *Model based learning and instruction in science.* Dordrecht: Springer.

Clement, J., & Steinberg, M. (2002). Step-wise evolution of models of electric circuits: A "learning-aloud" case study. *Journal of the Learning Sciences, 11*(4), 389–452.

Cohen, J. (1992). A power primer. *Psychological Bulletin, 112*, 155–159.

Dupin, J., & Joshua, S. (1989). Analogies and "modeling analogies" in teaching: Some examples in basic electricity. *Science Education, 73*, 207–224.

Engle, R. A., Conant, F. C., & Greeno, J. G. (2007). Progressive refinement of hypotheses in video-supported research. In Goldman, R., Pea, R. D., Barron, B., & Derry., S. (Eds.), *Video research in the learning sciences* (pp. 239–254). Mahwah, NJ: Erlbaum.

Hake, R. R. (1998). Interactive-engagement versus traditional methods: A six-thousand-student survey of mechanics test data for introductory physics courses. *American Journal of Physics, 66*(1), 64–74.

Hammer, D. (1995). Student inquiry in a physics class discussion. *Cognition and Instruction, 13*(3), 401–430.

Hestenes, D. (1987). Toward a modeling theory of physics instruction. *American Journal of Physics, 55*, 440–454.

Hogan, K., Nastasi, B. K., & Pressley, M. (2000). Discourse patterns and collaborative scientific reasoning in peer and teacher-guided discussions. *Cognition and Instruction, 17*(4), 379–432.

Hogan, K., & Pressley, M. (1997). Scaffolding scientific competencies within classroom communities of inquiry. In K. Hogan & M. Pressley (Eds.), *Scaffolding student learning: Instructional approaches and issues* (pp. 74–107). Cambridge, MA: Brookline Books.

Ingham, A. M., & Gilbert, J. K. (1991). The use of analogue models by students of chemistry at higher education level. *International Journal of Science Education, 13*, 193–202.

Keeley, P. (2008). *Science formative assessment.* Thousand Oaks, CA: Corwin Press and Arlington, VA: NSTA Press.

Khan, S. (2003). *Model construction processes in a chemistry class.* Paper presented at the National Association of Research in Science Teaching Conference. Philadelphia, PA, March 2003.

Korganci, N., Miron, C., Dafinei, A., & Antohe, S. (2015). The importance of inquiry-based learning on electric circuit models for conceptual understanding. *Procedia—Social and Behavioral Sciences. 191*, 2463–2468.

Lehesvuori, S., Viiri, J., Rasku-Puttonen, H., Moate, J., & Helaakoski, J. (2013). Visualizing communication structures in science classrooms: Tracing cumulativity in teacher-led whole class discussions. *Journal of Research in Science Teaching, 50*(8), 912–939.

McNeill, K. L., & Pimentel, D. S. (2010). Scientific discourse in three urban classrooms: The role of the teacher in engaging high school students in argumentation. *Science Education, 94*(2), 203–229.

Miles, M. B., & Huberman, A. M. (1994). *Qualitative data analysis: An expanded sourcebook* (2nd ed.). Newbury Park, CA: Sage Publications.

Minstrell, J., & Stimpson, V. (1996). A classroom environment for learning: Guiding students' reconstruction of understanding and reasoning. In R. Glaser & L. Schauble (Eds.), *Innovations in learning: New environments for education,*. Mahwah, NJ: LEA.

NGSS Lead States. (2013). *Next generation science standards: For states, by states.* Washington, DC: The National Academies Press.

Schwarz, C. V., Reiser, B. J., Davis, E. A., Kenyon, L., Achér, A., Fortus, D., & Krajcik, J. (2009). Developing a learning progression for scientific modeling: Making scientific modeling accessible and meaningful for learners. *Journal of Research in Science Teaching, 46*(6), 632–654.

Scott, P. H., Mortimer, E. F., & Aguiar, O. G. (2006). The tension between authoritative and dialogic discourse: A fundamental characteristic of meaning making interactions in high school science lessons. *Science Education, 90*(4), 605–631.

Steinberg, M. S. & Wainwright, C. L. (1993). Using models to teach electricity—The CASTLE Project. *Physics Teacher, 31*(6), 353–57.

Steinberg, M. S., et al. (2004). *Electricity visualized: The CASTLE project student manual.* Roseville, CA: PASCO Scientific.

Steinberg, M. (2008). Target model sequence and critical learning pathway for an electricity curriculum based on model evolution. In J. Clement & M. A. Rea-Ramirez (Eds.) *Model based learning and instruction in science* (pp. 79–102). Dordrecht: Springer.

van Zee, E., & Minstrell, J. (1997). Reflective discourse: Developing shared understandings in a physics classroom. *International Journal of Science Education, 19*, 209–228.

Vygotsky, L. (1962). *Thought and Language*, (E. Hanfmann & G. Vaka, trans). Cambridge, MA: MIT Press.

Wells, M., Hestenes, D., & Swackhammer, G. (1995). A modeling method for high school physics instruction. *American Journal of Physics. 63*, 606–619.

Williams, E. G. (2011). *Fostering high school physics students' construction of explanatory mental models for electricity: Identifying and describing whole-class discussion-based teaching strategies.* (Unpublished doctoral dissertation). University of Massachusetts, Amherst.

Williams, G., & Clement, J. (2015). Identifying multiple levels of discussion-based teaching strategies for constructing scientific models. *International Journal of Science Education, 37*(1), 82–107.

Windschitl, M., Thompson, J., & Braaten, M. (2008), Beyond the scientific method: Model-based inquiry as a new paradigm of preference for school science investigations. *Science Education., 92*, 941–967.

CHAPTER 5

EXTENDING THE BOUNDARIES OF HIGH-SCHOOL PHYSICS

Introducing Computational Modeling of Complex Systems

**Elon Langbeheim, Haim Edri, Nava Schulmann,
Samuel Safran, and Edit Yerushalmi**

This chapter describes a high-school level program in which students use and develop computational models of complex chemical or biological systems. Learning computer programming together with the physics principles that come into play in modeling complex systems, is a challenging undertaking. Thus, the introduction of programming in the program was designed gradually; from interpreting the physical meaning of a computer code, through modifying a familiar code, to constructing a novel computational model. Our analysis reveals that while most students were capable of interpreting a familiar code, only half of them were able to modify it properly. We also found that prior programming knowledge strongly influences students' ability to build their own computational models. Drawing on student responses to a feedback survey, we relate students' attitudes towards learning in the program to their performance in it. Finally, we reflect on our own experience as teachers and curriculum developers who introduce advanced physics topics using computational tools.

Keywords: Programming, computer simulations, student motivation, high school physics, conceptual knowledge, complex systems

Physics Teaching and Learning: Challenging the Paradigm, pages 111–133.

Physicists in academia and industry often apply the concepts and techniques of physics to investigate a wide range of systems in chemistry, biology and geoscience. Yet, in secondary school, physics is usually introduced in the context of centuries-old systems such as pulleys, springs and inclined planes. While these examples are necessary for introducing the fundamentals of physics, their application promotes a narrow and somewhat outdated representation of its scope. Such a representation of physics can discourage talented, high-achieving students who seek excitement, contemporary contact and inspiration in science.

Reports on special programs introducing contemporary physics topics at the secondary school level are sparse. For example, Etkina, Matilsky and Lawrence (2003) report a four-week summer astronomy program in which students use and interpret cosmological observations at a university setting. The program then continues with independent astronomy research projects that students conducted during the academic year. Langbeheim, Safran, and Yerushalmi (2016), describe an afternoon program for talented high school students which includes theoretical and experimental inquiry projects concerning the physical properties of soft materials—a topic usually studied at the graduate level.

This chapter describes the design and implementation of an Interdisciplinary Computational Physical Science (ICPS) program for excelling students, which takes place after school at a science education center. The crux of this program is a three-year long, inquiry-based learning curriculum, in which students study and develop computational models of multi-particle systems of matter. Students' final computational inquiry projects are credited with a matriculation grade appearing on their transcripts. Computational modeling of physical processes and phenomena is an aspect of current physics practice only marginally reflected in school. It encompasses the conceptualization, realization and analysis of computer programs mimicking the behavior of an actual system. The development of models requires simplifications of the intricate interactions in the real system, and theory based algorithms, that compute the dynamic or equilibrium properties of the simulated system.

Exposing students to relatively advanced topics with computational modeling as a major tool clearly challenges the common paradigm in physics education, which makes little use of computation. But, is it feasible? What do students actually learn in such a radically different program? Does their participation cultivate motivation to learn science? Our chapter will provide initial findings addressing these questions.

DEVELOPING AND USING COMPUTER MODELS FOR LEARNING SCIENCE

Despite recommendations by prominent science and mathematics educators (Chabay & Sherwood, 2008; diSessa, 2001), integrating computational modeling into physics curricula is not a widespread practice at the undergraduate level (Caballero, Kohlmyer & Schatz, 2012; Landau, Manuel & Bordeianu, 2015), and occurs even less at the high school level (Aiken et al., 2013; Taub et al., 2015). Most attempts at using computational modeling for learning physics at the high school and introductory college level focus on Newtonian motion problems with one or two interacting objects. Quantitative modeling of interacting, many-body systems, which comprise most biological and chemical systems, is very rare at the secondary school level (examples can be found in Wilensky, 2003; Wilensky & Reismann, 2006).

Computational modeling of complex systems in high school is a challenging task, but understanding the algorithms underlying the models of these systems and adjusting them is within the reach of students. For example, Wilensky & Reismann, (2006) demonstrated how high school students study biological phenomena such as the synchronized flashing of fireflies by building and modifying computer simulations in NetLogo (Wilensky, 1999). The students read about the phenomenon, and created a model in which each modeled firefly followed a set of rules of motion and interactions, that, at some density of fireflies would yield the synchronized flashing phenomenon. This shows that computational models allow high school students to study complex systems with mathematical models that are beyond their level (Wilensky & Reismann, 2006). Moreover, students engaged in theoretical inquiry projects who get only a partial handle on the mathematical models involved, might feel that their work is insignificant (Langbeheim, Safran, & Yerushalmi, 2016). This experience is quite different from students whose inquiry is mainly experimental (i.e. the students' major effort is to build an experimental setup and make measurements) and not the development of mathematical models. When working on experimental projects in high school physics, students feel ownership and satisfaction with their research (Kapon, 2016). This feeling of ownership and achievement is probably an important factor in the increase in motivation to learn science of students who attended science enrichment programs at academic settings (Stake & Mares, 2005). Similarly, students who build and use computational investigations of complex systems do not express lack of ownership in their research experience, even though it is theoretical (Langbeheim, Safran, & Yerushalmi, 2016). Thus, computational modeling can be a useful path for introducing advanced physics topics to talented students, thereby cultivating their interest and motivation.

Computational modeling encompasses several practices, such as using models to understand a concept, assessing or evaluating the consistency of models and constructing novel computational models (Weintrop et al., 2016). The simplest form of computational modeling –using computer simulations for developing conceptual understanding—is a widespread practice in education (Levy & Wilensky, 2009; Moore et al., 2014). The more challenging forms of computational modeling—assessing computational models by evaluating the consistency of the code with the physics of the modeled system (Weatherford, 2011) and programming novel computer simulations (Wilensky & Reismann, 2006) occurs mainly in special educational settings.

Thus, the practice of teaching students to construct computational scientific models is at nascent stage, and the literature on conveying programming skills needed for these models is scarce. One common method for teaching programming is using incomplete computer programs and asking students to complete or modify them (Van Merriënboer, & De Croock, 1992; Weatherford, 2011). However, reading and interpreting the text of a computer code without running the code and debugging it, does not guaranty the acquisition of the basic tenets of programming (Winslow, 1996). Actual programming on computers is a complex process of moving back and forth between writing the code, running it, and checking for errors. Therefore, it is not clear how reading and modifying code on paper is transferred to the capability of building the code on the computer (Winslow, 1996).

The core of computational modeling is translating a conceptual scientific model to a functioning computational procedure. Thus, building computational models that simulate natural systems, requires prior acquaintance with the conceptual model of the system and can then improve conceptual knowledge. (diSessa, 2001; Papert, 1980; Sherin, 2001). This can be achieved by first teaching the conceptual model without computation, learning the code features of the programming environment, and then letting the students construct a computational version of the model (Xiang & Passmore, 2015). Then, engagement in building computational models strengthens students' conceptual understanding of the system. For example, high school students develop their conceptual understanding of circular motion through building and modifying a computer code that simulates this type of motion (Taub et al., 2015). However, there is little evidence that introducing scientific models together with programming, actually improves students' content knowledge and skills. For example, Psycharis & Kallia (2017) found that mathematical problem solving performance of students who learned computer programming in conjunction with mathematics concepts, was not significantly different from the performance of the control group that did not use programming.

To summarize, although the effect of programming on learning standard physics topics is unclear, it unquestionably enables students to model advanced systems that would not be accessible to them otherwise. Since computational science models requires an understanding of programming, teaching computational science concerns the development of both programming competency and conceptual understanding of scientific content. In the following, we explain how we implemented such a dual approach in a computational science program for high school students.

CONTEXT OF THE STUDY

The Interdisciplinary Computational Physical Science (ICPS) program takes place once a week in the afternoon at a regional science education center in Israel. The goals of the program are to introduce students to contemporary topics in physics, to reflect the central role of computation in contemporary science, and to cultivate talented students' motivation to pursue a career in science. In order to participate in the program, students must take an advanced science course (equivalent to A-Level in the UK and AP in the US) at their school and pass an entrance exam that assesses logical reasoning skills. The program is three-years long (10th to 12th grade) and is built on an inquiry-based learning curriculum in which students develop computational skills through modeling tasks (Taub et al., 2015) and conduct comprehensive inquiry projects.

The ICPS program introduces students to the physical and computational foundations for modeling complex physical systems that are important in physics, chemistry, materials science and biology. The program includes three major stages in building these competencies, covering 10th–12th grades respectively. The focus of the 10th grade curriculum is the analysis of Brownian motion and diffusion via computational models of elastic spheres. A central theme in this first unit is bridging between deterministic Newtonian models and stochastic, random walk models. The stochastic dynamics of the individual particles paves the way to the 11th grade curriculum focusing on introductory level statistical thermodynamics, which includes the ideas of entropy, heat and internal energy. The focus of the 12th grade curriculum is the analysis of complex structure formation in materials, using Monte Carlo computational methods. This unit is adapted from an earlier and more limited curriculum for talented high school science students (Langbeheim, Livne, Safran, & Yerushalmi, 2012; Langbeheim, Safran, & Yerushalmi 2016).

The computational models studied in the program were developed using the VPython package (Scherer, Dubois, & Sherwood, 2000) containing a library of three-dimensional graphical objects that can be used in simulations. The package was designed for the Python programming language and has become the language of choice for educational computational physics (Caballero, Kohlmyer, & Schatz, 2012; Chabay & Sherwood, 2008; Landau, Manuel, & Bordeianu, 2015). The VPython code is written in an Integrated Development system called IDLE that organizes the formatting of the program. IDLE requires the user to structure the code according to a hierarchical format where indentation represents an inner loop, a function or an if-else condition.

OUTLINE OF THE STUDY

In the following sections, we elaborate some parts of the ICPS program curriculum in more detail, show what students do and accomplish in different stages of the program, and how they perceive the ways of learning in the program. The first section examines students' comprehension of computational models and their ability to assign physical meaning to the simulation code at a relatively early stage of the program—the middle of the 10th grade. In the subsequent question, we examine students' construction of computational models in the last stage of the program—at the end of the 12th grade. Then, we discuss the role of teachers in guiding students' in constructing computational models. Finally, we examine students' experiences of learning in the program. The description is guided by the following research questions:

a. How do students in the program interpret and modify the code of computational models?
b. What factors facilitate or limit students' capability to build computational models on their own?
c. What characterizes teachers' involvement in the students' computational projects in the program?
d. How do students describe their learning experience in the program and its influence on their motivation to pursue a career in science?

The population is approximately 13–17 students in each grade level with 47 students in all three cohorts at the time of this study. Students can only join the program at its outset in 10th grade (after passing initial screening), and few leave during the program so that the 10th grade cohort is usually larger than the 11th and 12th grade cohorts. We use a qualitative methodology and multiple sources of data such as students' responses to midterm exams, emails, paper drafts, versions of simulation code and a feedback survey, in order to address these research questions.

INTERPRETING AND MODIFYING COMPUTATIONAL MODELS OF PHYSICAL PROCESSES—10TH GRADE

The main themes in the 10th grade instructional sequence in the ICPS program are the phenomena of diffusion and Brownian motion. These phenomena are studied using systems of particles with repulsive, short-range interactions, and realized using computational models based on the Molecular Dynamics (MD) method. A common MD model for diffusion is the "elastic sphere gas" model, which is based on a simple approximation; particles interact through elastic, short-range radial interactions. By estimating the interaction using Hooke's law, the algorithm calculates the acceleration of each sphere using Newton's second law of motion. Given the acceleration, the Euler integration method (Scherer, 2010), is used to calculate the velocity and position of each particle. Euler integration is a simple and somewhat inaccurate method. More sophisticated methods such as the Verlet scheme or the Leap-Frog scheme (Frenkel & Smith, 2001) are more accurate, but also more complex and not necessary for simulating the elastic sphere gas.

The basic elements of the MD simulation for the elastic sphere gas were introduced in the 10th grade at the beginning of the program. The MD method was introduced using seven scenarios with increasing complexity that reflect an element in the particles' motion before, during and after they collide with each other and with the container walls. These scenarios, shown in Figure 5.1, are organized by the dimensionality and the number of particles involved, and ordered from the simplest situation of constant velocity motion (rubric 1) to the most complex

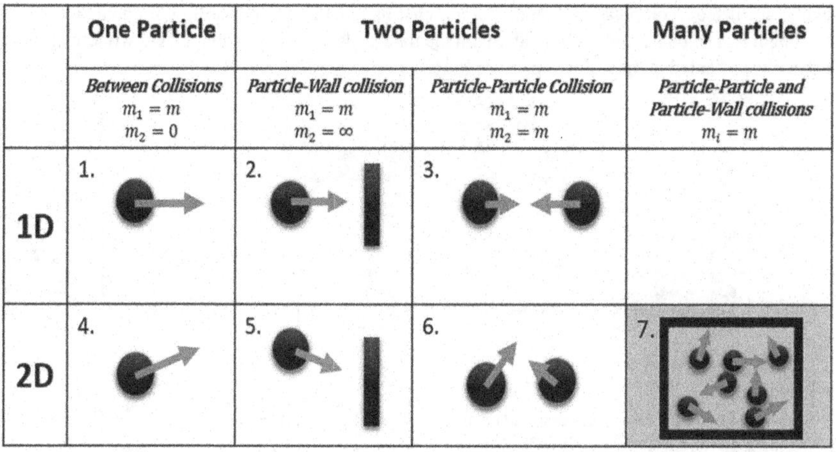

	One Particle		Two Particles		Many Particles
	Between Collisions $m_1 = m$ $m_2 = 0$	*Particle-Wall collision* $m_1 = m$ $m_2 = \infty$	*Particle-Particle Collision* $m_1 = m$ $m_2 = m$		*Particle-Particle and Particle-Wall collisions* $m_i = m$
1D	1.	2.	3.		
2D	4.	5.	6.		7.

FIGURE 5.1. The seven scenarios of the elastic-sphere gas used as a framework to learn to model the different elements of motion in diffusion of gases.

situation of many particles that collide with each other and with the walls (rubric 7). Some of these scenarios were introduced using experiments in which students measured and represented the velocity and position of carts colliding with each other or with a wall.

The first research question asks how students interpret and modify the code of computational models after the initial stage of the program? Specifically, we are focusing on whether students are able to identify the physical concepts underlying the Molecular dynamics model of the elastic-sphere gas and to modify it adequately. We use student responses from an exam administered at the end of the first semester as data. We focus on two open-ended tasks about code interpretation and modification shown in Figures 5.2 and 5.3.

The task in Figure 5.2 asks students to assign each line of code to a cluster, and to explain the function of each cluster of code. The canonical MD code includes three main clusters: a). initialization, b). force calculation and c). the integration of the equations of motion (see for example Frenkel & Smith, 2001, pp 64–74). The initialization includes the assignment of initial values to parameters such as the spring constant, the length of the arena (or the position of the walls), and the initial positions and velocities of the particles (lines 1–7 in Figure 5.2). When producing a visual output, the initialization includes also the creation of graphical objects such as the ball in (lines 8–9). The main loop includes the force calculation—in our case a repulsive spring force when colliding with either the left or right wall (lines 12–15), and the motion is updated according to a different cluster, in this case, the simple Euler integration method (lines 16–19).

The analysis of student responses was carried out in two steps. In the first, students' cluster assignments were compared to the expected, canonical clusters

David constructed a computational model for a rubber ball that moves in a hollow frictionless tube with length L. The ball collides elastically with the tube's edges. The code that David wrote with VPython is presented in the following:

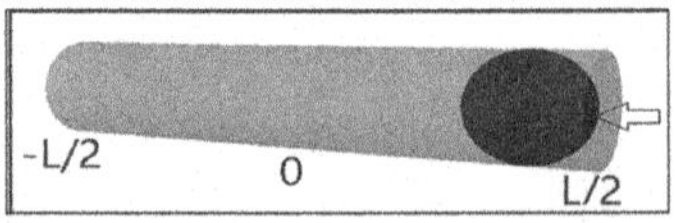

```
                             -L/2        0               L/2
1. m = 2.0
2. k = 100.0
3. R = 0.5
4. L = 10.0
5. dt = 0.001
6. v = vector(5.0,0,0)
7. tube = cylinder(pos=vector(-L/2,0,0), axis=(L,0,0), radius=R)
8. ball = sphere(pos=(-3.0,0,0), radius = R)

9. while t < 10:
10.     rate(100)

11.     if ball.pos.x >  L/2: F1 = -k*(ball.pos.x-L)*vector(-1,0,0)
12.     else: F1 = vector(0,0,0)
13.     if ball.pos.x < -L/2: F2 = -k*(ball.pos.x-L)*vector(1,0,0)
14.     else: F2 = vector(0,0,0)

15.     F_net = F1 + F2
16.     a = F_net/m
17.     v = v + a*dt
18.     ball.pos = ball.pos + v*dt

19.     t = t + dt
```

Divide the code into clusters according to their role in the computational model (a cluster does not have to consist of adjacent lines). Describe in short the role of each cluster and elaborate on the meaning of the lines. Use the following table and the example shown in its first line.

Cluster	Role of the Cluster	Meaning of the Lines
1-9	Initialization Defining the system's parameters	M – ball's mass, k – elastic constant, R – ball's radius, L – tubes length

FIGURE 5.2. The task from the midterm exam which asks students to divide the code into clusters, describe the role of each cluster and interpret the meaning of code lines.

of the MD model. In the second, students' attributed meaning of the clusters was sorted into two main categories: physical or computational. To that end, we categorized students' explanations of the role of each cluster as either physical and/or computational. The unit of analysis usually includes a sentence or a set of several sentences, which demonstrate a single idea.

Table 5.1 illustrates two students' answers for the definition and meaning of code lines in cluster B—force calculation. The description of the cluster given

TABLE 5.1. Sorting Two Students' Clusters to Physical or Computational Category in Cluster B: Force Calculations

	Lines	Cluster's Role	Meaning of the lines	Category
Expected answer	12–16	Force calculation for short range repulsive interactions	Each if-else pair of lines (12,13 and 14,15) calculates the forces during the ball's collision with each wall: if the ball's position is greater/ smaller than the right/left wall's position, the force acts in $-\hat{x}/+\hat{x}$ direction (respectively) and its magnitude is calculated using Hooke's law. If the conditions do not meet these criteria, the force equals zero. Line 16 combines F1 and F2 into a single entity—the Net force	Physical
S6 answer	12,14	Responsible for collision	Responsible for determining the force during the collision using Hooke's Law	Physical
	13,15		Determines the forces on the balls not during the collision (No force)	
	16		Adding the forces in order to find the resultant force on the ball	
S12 answer	12–15	Defining the location in which the force will act on the ball	If – something happens – else – if it does not	Computational

by student S6 is consistent with the expected categorization of the force calculation cluster. It has the relevant physical concepts (Hooke's law, calculation of net force) and reflects understanding of the physical meaning of the force calculation. Similarly, the description given by student S12 indicates that this cluster in the code expresses a condition for the force; however, when addressing the meaning of each line of the code, he assigned mostly computational interpretations and not the physical ones.

Most students identified the three main clusters: initialization, force calculation and integration of the equation of motion (13/17). They grouped lines 12–15 as a separate cluster, similar to the expected answer. In their descriptions of the clusters, most of the students used concepts and terms of kinematics (acceleration, velocity, position) and of Newtonian dynamics (force, net force, Newton's laws) to describe the different parts of the code, but did not use the appropriate technical terms such as equation of motion, initial conditions or the Euler integration method. The following description illustrates a common response: "According to the force and the mass (of the particles) we can calculate the acceleration—from Newton's 2nd law. The velocity can be calculated according to the acceleration and the position can be calculated according to the velocity." Nine of the 13 students

Suzi suggests replacing a few lines of David's code (see Figure 5.2) with the following lines:

```
if ball.pos.x < -L/2: v.x = -v.x
if ball.pos.x >  L/2: v.x = -v.x
                  \
```

1.) Which lines in David's code can be replaced with the above suggestion?
2.) What is the function of these lines in David's code?
3.) Why can they be replaced with Suzi's lines?

FIGURE 5.3.   The second task in the midterm exam, asking about the modification of the code.

who made the correct grouping of lines in the code, identified the physical meaning of the cluster whereas four focused mostly on its computational aspects.

The second task from the midterm exam asked students about the modification of the code from Figure 5.2. The task presented two lines of code that can replace the force calculation and the Euler integration method with a simple velocity inversion rule shown in Figure 5.3:

The task required students to identify and compare the main elements in David and Suzi's models, and to state which lines in David's code should be omitted when using Suzi's velocity inversion rule. Students' suggestions for replacing code lines were analyzed using three categories: 1.) full agreement with the expected, expert-level answer— replacing lines 1,2, 12–18 or lines 12–18 in David's code,

TABLE 5.2.   Categorization of Responses to Question 2, Function of Lines: Expected Answer, Two Students' Responses and the Associated Categorization

	Lines to be replaced	Category	Justification for the replacement of the lines	Category
Teacher	1, 2, 12 - 18	Full	"Suzi's suggestion models the result of the ball-wall elastic interaction of David's model. According to Suzi's suggestion, the ball's position is calculated only before and after the collision, based on phenomenological flipping velocity rule. Therefore, all lines that update force, acceleration and velocity in David's model can be replaced with Suzi's suggestion."	Full
S17	12 - 18	Full	"The role of these lines is to activate the force when the ball reaches its end and creates a collision. When this happens, there is also the acceleration, and then there is a change. They can be replaced because of what the Suzi's suggests is also true, and the ball will change its direction even if we use Suzi's suggestion."	Full
S2	12,14	No agreement	"These lines describe the condition in a more organized manner than in David's code, but in less precision."	No agreement

2.) partial agreement with the expected answer—replacing only lines 12–15 or lines 16–18 in David's code, and 3.) no agreement with the expected answer—other suggestions for replacement. Table 5.2 shows the response of S17 which in full agreement with the expected response and that of S2 which gives no relevant justification of the replacement.

Students' explanations for the replacement of the code lines were analyzed vis-a-vis the expected answers: a.) using physical terms such as elastic interaction, acceleration and velocity, to distinguish between the model's rules and its predictions, b.) partial explanation of the model's rules and poor description of the model's predictions, c.) explanation is missing, wrong or unclear.

Figure 5.4 shows the proportion of responses in each category of the code modification task. Nearly half of the students (8/17) responded correctly that the force calculation and the part of the Euler integration algorithm (which updates the acceleration and the velocity) should be replaced. About a third of the students (6/17) suggested replacing only the force calculation cluster, missing the fact that updating the acceleration and the velocity is no longer necessary when using the velocity inversion rule to model the collision and another student suggested replacing only the cluster of updating motion from interaction (lines 16–18 in David's code). The answers of two students were considered in no agreement with the expected response: they proposed replacing only the "if" part of the conditioning sentence within the force calculation cluster. When asked to explain their suggestions, less than half of the students (7/17) managed to provide the appropriate explanation, namely, that both models produce the same behavior in the system: particles bounce of the wall with the same speed in the opposite direction. The rest were either partial or irrelevant explanations.

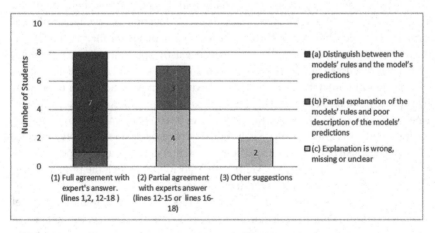

FIGURE 5.4. Responses to the second task asking which lines in David's code (shown in Figure 5.2) can be replaced with the velocity inversion rule (shown in Figure 5.3)?

To conclude, in this task, students were asked to compare two computational models for a particle-wall collision: the Euler integration of the spring-like force calculation and the velocity inversion. The task required deeper understanding of the physical meaning of the code than the first task. Our analysis of student responses indicates that almost all students suggested to remove the force calculation cluster, but only half of the students suggested to also remove the lines updating the acceleration and the velocity. The comparison of students' performance on the task in Figure 5.2 and on the task in Figure 5.3 indicates that modifying the code is a more challenging task than interpreting a given code. While the vast majority of students identified the clusters of code in the program correctly, only about half of the students identified the parts of the code that should be replaced correctly, and even fewer students made appropriate justifications for the replacement.

CONSTRUCTION OF NOVEL COMPUTATIONAL MODELS OF COMPLEX PHENOMENA—12TH GRADE

The 10th and 11th grades in the program focus on basic theoretical and computational elements of modeling multi-particle systems. The 12th grade curriculum continues to the next level: modeling the emergence of structures in interacting multi-particle systems using the Monte-Carlo simulation method. In this method, the behavior of a system is simulated by sampling configurations of the system, generated using random numbers. The method can be realized through sampling independent configurations of the system, but this is not common. The more common method of studying dense systems of interacting particles is the Metropolis algorithm (Krauth, 2006). This algorithm samples the configurations of the system by starting from an arbitrary configuration, and altering the state of the system using small, random steps that change the location or orientation of each particle. The acceptance probability of each step depends on the interactions in the systems. Take, for example, a system of elastic spheres, whose Newtonian model was discussed in the previous section. In such a system, a sphere may move at each step in a random direction. If, however, the move leads to an overlap of two spheres; the move is rejected. This means steps are not entirely random, and when restrictions on steps are frequent, an ordered, non-random configuration emerges in the system (Krauth). Such rigid objects models do not involve interaction energy considerations. Inter-particle interaction energies and temperature are brought into the arena using the Metropolis algorithm in which the probability of steps is weighted by the Boltzmann factor (Krauth,).

The 2nd part of the school year in the 12th grade, is dedicated to building novel computational models in research projects. Students investigate phenomena in systems of matter, and build computational models that utilize the Monte-Carlo method. We describe the challenges students faced and their interactions with the teacher at this stage. Specifically, we will examine the second research question listed above: what factors facilitate or limit students' capability to build computational models on their own?

Utilizing a qualitative case study approach, we focus on two student pairs conducting their final research project in the 12th grade. These students came from a cohort of 13 students who started the program a couple of years before the 10th graders described in the previous section. We selected two pairs of students whose performance on the midterm exam was similar to the class average. The midterm exam in the 12th grade consisted of questions in which students were asked to interpret lines of code and to explain their physical meaning similar to the examination given in the 10th grade. As we will show, despite the similarity in their performance on the exam, the enactment of the computational projects was very different in each pair. We use this comparison to highlight the mismatch between students' learning outcomes in the paper and pencil exam, and their ability to develop computational models of complex phenomena.

The first pair were two female students, Nataly and Leah (pseudonyms), both of whom were highly committed to the program and had prior experience in programming. Nataly and Leah wanted to study the coagulation of blood, but since this process is very complex, the teacher suggested a simpler, similar process; the agglutination of blood. Agglutination is the clumping of blood cells when blood transfusions of the wrong blood group X react with the antibodies of the host blood type Y. The antibodies of blood type Y attach to the X blood cells in order to stick them together and screen them out of the system. The antibodies attach to sites on the blood cells that are called antigens and are shown in the model in Figure 5.5 below.

Nataly and Leah studied the phenomenon in depth, reading popular explanations on the internet, as well as primary research literature. They constructed a

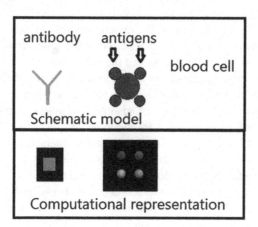

FIGURE 5.5. The model of blood cells and antibodies in a sketch (top) and in the simulation (bottom)

model, which was an extension of a lattice-based simulation of a two-component mixture that was studied in class. Contrary to the lattice-mixture model which has two types of particles of the same size, their model included two entities (antibodies and blood cells) with different sizes and binding geometries. They developed several functions in the program on their own, modeling the system on a square lattice (see Figure 5.5). However, their final program did not work exactly as they intended. They planned to limit the number of blood cells that can clump together with one antibody, because in reality, each antibody has only two binding sites. Limiting the number of neighboring interactions required a complex function. Because the deadline for submission was approaching, the teacher suggested that this function is not necessary. He explained that due to the square lattice geometry, a configuration in which one antibody has three neighboring blood cells is not very likely (due to the obstruction created by the large blood cells) and thus a function that limits the amount of interactions would not alter the simulation results dramatically. After they replaced the complex function with a simpler one, Leah and Nataly ran the simulation and produced results for different proportions of blood cells and antibodies. They wrote a comprehensive, 32 page long report on their research and submitted it on time.

The second pair were two male students: Roy and Guy,—who, in contrast to Leah and Nataly, had no prior experience in computer programming. Roy and Guy were also committed to the program, but were quieter and seemed less engaged in class discussions than Nataly and Leah. At the beginning of the project period, they were asked to suggest a topic for inquiry. As the teacher noticed that they could not come up with an appropriate topic, he suggested they investigate the models for adsorption of a water-dissolved solute to a solid surface (Girado & Ayllo, 2011). In order to gain some insight on the phenomenon, the teacher suggested an experiment in which they would use water with varying concentrations of food coloring as the adsorbent and a small amount of activated charcoal as the adsorbing surface. The purpose of the experiment was to examine the volume fraction of the adsorbed material as a function of the concentration of the adsorbent in the solution. The relation between the concentration of the solute and the fraction of the adsorbed material can be described using two theoretical models: the Langmuir model, which assumes a single-layer adsorption, and the Freundlich model that assumes multiple layers of adsorption (Girado & Ayllo, 2011). Roy and Guy conducted the experiment for the most part on their own, sporadically consulting the teacher when their results seemed problematic. Next, they started working on a computational model that simulates the adsorption process. At this point, they seemed to be at a loss and consulted the teacher. The teacher told them to think about how to model the interaction energy between the surface and the adsorbent particles. The teacher expected them to modify the simulation of the two-component mixture which they studied in class, where the internal energy of system changes as particles of each component attach or detach from particles of the other component. In the modified simulation, particles would lower the inter-

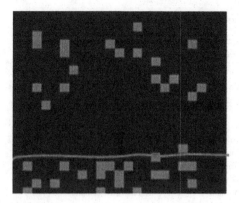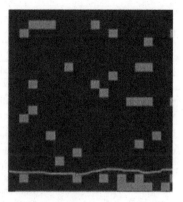

FIGURE 5.6. The model at the left has four adsorption layers and the one on the right has two layers. The thin line marks the separation of the top layer from the bulk fluid.

nal energy of the system when they attach to the surface and increase the energy when they detach. However, Roy and Guy could not implement this modification on their own. The teacher had to guide them through the modification process step by step. Eventually, they produced a working model, with a graphic output showing the solute and the adsorbed material as shown in Figure 5.6.

After running the model several times in order to calculate the average fraction of adsorbed particles, Guy and Roy discovered their model was producing problematic results. Again, they were not able to figure out the problem on their own. The problem stemmed from an error in the calculation of the density of the adsorbed particles, and the teacher had to discover and fix the error. Guy and Roy who had little programming experience, were not able to debug the program and were mostly occupied in running the simulation in different conditions (temperature, initial concentration) or changing its graphics. They summarized their results in a final paper describing the function of the model, and the results of the amount adsorbed vs. concentration of solute. However, they failed to produce an accurate comparison of their results with the mathematical model, and their written work was much shorter and less comprehensive than Nataly and Leah's report.

The comparison reveals vast differences between the two pairs: Nataly and Leah's project required complicated modeling considerations and a sophisticated computer code. Nevertheless, they planned and managed the programming mostly on their own (although they needed the teacher's help at some point). Roy and Guy's model required minor modifications, but they were not able to produce them on their own. Leah and Nataly's code comprised twice the number of lines of code, and their paper was significantly longer and more comprehensive than was Roy and Guy's paper.

TABLE 5.3. The Main Differences Between The Two Pairs in Their Final Computational Projects

Feature of the Project	Nataly and Leah	Roy and Guy
Source of the idea for research	The students came with an initial idea, the teacher provided focus	The teacher gave the idea
Students' independence in writing the code	High, although not completely independent	Low, most of the code was given by the teacher
Length of the paper	32 pages	18 pages
Complexity of the model	High, the computational implementation of the model required extensive modifications from and additions to the familiar code. Its length was 180 lines of code.	Low, the model required only minor modifications of the original, familiar code. Its length was 89 lines of code.

To summarize, the level of independence and depth of the projects differed greatly, although student performance on the midterm exam was similar and both students pairs were finishing three years in the program. The differences in performance could stem from character differences, gender and other factors. However, in our experience, usually female students are not more independent or hard working than the male students. We therefore attribute at least part of the difference to the very different levels of prior programming knowledge the students had. Nataly and Leah's prior programming knowledge enabled them to progress more independently than Roy and Guy in constructing the computational model. This reflects the group as whole, indeed students with prior programming experience are more independent in their work. Of the 14 pairs that graduated the program at the time this chapter was written, only two pairs with no computer science background worked independently on their projects. Four pairs with prior programming experience like Leah and Nataly, built their computational models with little help from the teacher, and the rest (8 out of 14) needed extensive help from the teacher like Guy and Roy. Other factors such as intrinsic interest and motivation also contributed to the difference in the scope and complexity of the projects. As we show later on, the difference between the two pairs revealed itself not only in their performance in the final project, but also in their comments on learning in the program in general.

TEACHER INVOLVEMENT IN STUDENTS' COMPUTATIONAL PROJECTS

In this section, we address the third research question regarding the involvement of the teachers in the students' computational projects. Four of the authors of this chapter were teachers and curriculum developers in the program (Langbeheim, Edri, Yerushalmi, & Schulmann). Because we developed many of the activities and simulations on our own, we often faced challenges resembling the challenges students face when they build computational models. Nevertheless, as experienced physics teachers who took university level courses in programming, we were able to judge our models' self-consistency and the results they produced vis-à-vis results that were reported in textbooks and academic articles.

As we have shown in this chapter, developing computational knowledge alongside physical content is a challenging task at the edge of students' abilities. In order to analyze learning at the edge of students' abilities, we apply Wertsch's (1984) interpretation of the Zone of Proximal Development (ZPD). The Zone of Proximal Development is a term suggested by Vygotsky (1978) to describe learning of topics beyond the developmental level of students. The complexity of these learning tasks prevents the learners from progressing independently, but they may be able to do so under adult-expert guidance (Wertsch). Similarly, students in the ICPS program were novices in the field of computational science and needed the help of a teacher for building their projects. Wertsch demonstrates the process in which a competent expert guides a mentee, using an example of an adult who teaches a child how to build a jigsaw puzzle. The expert-adult knows it would be much easier to use the picture of the complete puzzle for guidance, whereas the child does not have a clear strategy for doing the task. Thus, experts have certain heuristics that help them to structure tasks in order to complete them, although they do not necessarily know exactly what to do at each step. Similarly, the teachers in the ICPS program have a wider repertoire of ideas and resources than their students, although they do not necessarily have a clear blueprint of the simulation the students strive to build.

Working with students in the Zone of Proximal Development requires the teachers to elicit the ways students understand the tasks. After understanding the task from the students' perspectives, the teachers can think about ways which can link the students' perspective of the task to their own understanding. If, for example, students say they cannot understand a complex part of a code, the teacher can read this part of the code line by line with them, asking the students to write down the meaning of each line, and to suggest how to combine lines that are part of a distinct procedure into a cluster.

Since learning physics alongside programming is not common practice, pedagogical suggestions for teachers are scarce. We believe teachers who plan to teach a course that includes programming need to be proficient in reading code of programs and developing at least parts of the computer programs on their own. In addition, they need to find ways to understand the students' challenges in the computational tasks they are doing. This expertise, much like addressing students' misconceived ideas in physics, comes with experience and through discussions with colleagues who engage in similar endeavors.

STUDENTS' PERCEPTIONS OF LEARNING INTERDISCIPLINARY COMPUTATIONAL PHYSICAL SCIENCE

The last research question addresses the affective aspects of learning in the ICPS program. We examine how do students describe their learning experience in the program and its influence on their motivation to pursue a career in science. In order to gauge students' learning experience in the program and its influence on their perceptions of scientific practice in the field, we administered an attitudinal survey at the end of the year (see appendix). Approximately two-thirds of the participants in the program (n=32/47) responded to the survey, as this was the end of the year and many were absent. Here we analyze three items from the survey: the first addressed the balance between science and computation in the program, the second addressed the difference in the ways of learning between school science lessons and the program and the third asked the students whether the program influenced their plans for learning in the future. We used the verbal analysis method (Chi, 1997) to identify common themes among student responses.

A majority of students (59%) who responded to the first item stated there was a proper balance between science and computation in the program, 22% stated that more computation is needed, and 16% stated that more science is needed and less computation. Nataly and Leah, who had prior experience in programming responded that the balance was appropriate, but Roy (who did not learn computer science in school) thought that more computation was needed (Guy did not complete the survey).

The second item asked how the ways of learning in the program differ from regular science learning in school. Reading through student responses reveals that several perceived differences were mentioned by students. Leah's response illustrates some of these categories:

> First, the topics that we learn are very different, the classes are smaller and there is more personal contact with the teachers. In addition, in the ICPS program we ask more questions than in school, the area of study is related to many phenomena that can be studied and we also acquire tools for inquiry. Therefore, inquiry and learning are not as limited as in school. The use of computers is also not constrained as it is in computer science courses in school. We learn the computational tools (in ICPS), but we are encouraged to think of many ways to use them when programming each problem.

Leah wrote that learning is more open-ended in the program and was conducted in smaller classrooms with a better atmosphere than in school. She also addresses the central role of computers in learning in the program and the encouragement of independent thinking and learning.

In the whole group of respondents, the most common theme, mentioned in 34% of the responses, referred to "independent/self-paced learning" as distinguishing between the ways of learning in the program and in school. Only 28% of the responses mentioned coding and the use of computers in the program as a major difference. Nataly and Leah both mentioned the open-endedness and the opportunities for independent study in their answers, whereas Roy only mentioned the projects and the experiments as a unique feature of the program but did not allude explicitly to independent learning nor to computation.

The third survey item asked students whether they think that the program influenced their plans for future studies. Of the students, 44% answered that the program positively influenced their plans to study science and/or computer science. For example, Nataly wrote: "The ICPS program helped me foster a more positive experience in learning science than school. Learning in school was a little boring... Had I learned only school science, I don't think I would have a positive experience and a motivation to learn a STEM related topic in the future." Conversely, 11 respondents (34%) wrote that the program did not affect their plans. Three of them added they planned to pursue a science career anyway and the program did not change their plans. Only one student wrote that the program had a negative influence on her plans. The rest wrote they have not decided what they will learn or did not respond to this question.

Leah, Nataly and Roy's responses to the survey second the claim we made before, namely, that the difference in the level of independence and scope of the project work of the two pairs is related to their prior knowledge in programming. This difference manifests itself in their perception of the balance of learning computation and science in the program: Roy responded that the computational component in the program is insufficient, whereas Nataly and Leah found it balanced. In addition, Nataly and Leah perceived the tasks as open-ended and mentioned the opportunities for independent study in the program when they compared learning in the program to school learning, whereas, Roy did not.

SUMMARY

Computational models, such as those investigated in the ICPS program, enable students with relatively little programming background knowledge to model complex physical systems (Wilensky & Reismann, 2006; Xiang & Passmore, 2015). The ICPS introduces computational physics models of complex systems to students, and students are then able to interpret the computer code of the model alongside its physical meaning. Similar findings were reported for college students who used computational models in an introductory physics course (Weatherford, 2011). We also found that modification and construction of code, which require a deeper and more holistic conceptualization of the computational procedures, were more challenging tasks for students. Less than half of the students (8/17) suggested modifying a familiar code properly. Similarly, the proportion of students who were able to modify an incomplete code in a college physics course was between 51% and 64% (Caballero, Kohlmyer, & Schatz, 2012). The analysis of two student pairs constructing their own computational models in the third year of the program indicates that prior programming knowledge is important for building computational models. We found that students with some programming background were able to construct computational models with little guidance, whereas many of the students who had no programming background were less confident in building the model on their own. Programming background is needed for constructing complex computational models in the ICPS program, but is not necessary for experimental or observational science projects such as the project in Etkina, Matilsky, & Lawrence, (2003) or Kapon (2016). Thus, students need prior knowledge in programming in order to construct computational science projects, much like they need prior knowledge in mathematics for solving physics problems.

We use the lens of ZPD to discuss the role of the teacher in modeling the construction of a computational model to the students. We argue that teachers need to develop the expertise of building computational models, but to also gain experience in uncovering the students' perspectives of their modeling tasks, in lieu of their immaturity in computer programming. Our experience resembles the descriptions of the teachers who used the Boxer programming environment: teachers adapt their instruction of computational modeling to the ways they understand and use the computational tool (diSessa, 2001). Getting to know the underlying algorithm of simulations can expand the teachers' understanding of the models they introduce (Wilensky, 2003).

Perhaps most importantly, we provide evidence that building computational models is a viable avenue for fostering the interest of talented high school students and cultivating their motivation to pursue a career in science. Almost half of the students in the ICPS program reported the program had a positive influence on their future study plans. The positive motivational effect of science enhancement programs on talented high school students has been reported before (Kapon, 2016; Stake & Mares, 2005). However, the source of the motivation was somewhat surprising. When the ICPS students were asked to explain how the ways of learning in the program differed from the ways of learning in school, many students wrote that learning ICPS was more open-ended and more independent. Interestingly, this distinguishing aspect was mentioned even more than the use of computation in the program, which is its main tool. In other words, inquiry-based computational modeling is a promising avenue for motivating talented students to explore contemporary physics, because it enables independent, open-ended exploration.

APPENDIX—ATTITUDINAL SURVEY ITEMS

1. Is the combination between computational parts and the scientific parts of the ICPS adequate?
   a. Yes
   b. No, the computational parts should be strengthened.
   c. No, the scientific parts should be strengthened.
   d. The computational and scientific parts should be combined differently

2. How are the ways of learning and working in the ICPS program different from learning and working in your science lessons in school?
3. Has learning in the ICPS program influenced your plans for future studies?
4. Does the science you learned in the ICPS program seem more "authentic" than the science you learned in school?
5. What, in your opinion, is the main contribution of learning in the ICPS program as related to learning computer science or science in school? (you may choose more than one option)
    a. Deepening the understanding of scientific and computational concepts that were learned in school
    b. Using computational tools to study a larger variety of natural phenomena and problems than those studied in school
    c. Learning new scientific concepts that I did not learn in school science lessons
    d. Deepening my understanding of computational concepts that I learned in the computer science lessons in school
    e. Applying tools that I acquired in computer science lessons in school to tackle scientific problems
    f. Learning new computational methods that were not used in computer science lessons in school
    g. Other: _____
6. Are science and computer science lessons in school important as a basis for the topics you studied in the ICPS program? Please explain

REFERENCES

Aiken, J. M., Caballero, M. D., Douglas, S. S., Burk, J. B., Scanlon, E. M., Thoms, B. D., & Schatz, M. F. (2013, January). Understanding student computational thinking with computational modeling. In *AIP Conference Proceedings, 1513*(1), 46–49).

Caballero, M. D., Kohlmyer, M. A., & Schatz, M. F. (2012). Implementing and assessing computational modeling in introductory mechanics. *Physical Review Special Topics-Physics Education Research, 8*(2), 020106.

Chabay, R., & Sherwood, B. (2008). Computational physics in the introductory calculus-based course. *American Journal of Physics, 76*(4), 307–313.

Chi, M. T. (1997). Quantifying qualitative analyses of verbal data: A practical guide. *The Journal of the Learning Sciences, 6*(3), 271–315.

DiSessa, A. A. (2001). *Changing minds: Computers, learning, and literacy.* Cambridge, MA: MIT Press.

Etkina, E., Matilsky, T., & Lawrence, M. (2003). Pushing to the edge: Rutgers astrophysics institute motivates talented high school students. *Journal of Research in Science Teaching, 40*(10), 958–985.

Frenkel, D., & Smit, B. (2002). *Understanding molecular simulation: From algorithms to applications* (Vol. 1). London, UK: Academic Press.

Girado, G.,& Ayllo, J. A. (2011). A simple adsorption experiment. *Journal of Chemical Education, 88,* 624–628.

Kapon, S. (2016). Doing research in school: Physics inquiry in the zone of proximal development. *Journal of Research in Science Teaching, 53*(8), 1172–1197.

Krauth, W. (2006). *Statistical mechanics: Algorithms and computations.* New York, NY: Oxford University Press.

Landau, R. H., Manuel, J. P., & Bordeianu, C. C. (2015). *Computational physics: Problem solving with Python.* New York, NY: John Wiley & Sons.

Langbeheim, E., Livne, S., Safran, S. A., & Yerushalmi E. (2012). Introductory physics going soft. *American Journal of Physics, 80,* 51–60.

Langbeheim, E., Safran, S. A., & Yerushalmi, E. (2016). Scaffolding students' engagement in modeling in research apprenticeships for talented high school students. In Keith, S. Taber & Manabu Sumida (Eds.), *International perspectives on science education for the gifted: Key issues and challenges.* New York, NY: Routledge.

Levy, S. T., & Wilensky, U. (2009). Students' learning with the Connected Chemistry (CC1) curriculum: Navigating the complexities of the particulate world. *Journal of Science Education and Technology, 18*(3), 243–254.

Moore, E. B., Chamberlain, J. M., Parson, R., & Perkins, K. K. (2014). PhET interactive simulations: Transformative tools for teaching chemistry. *Journal of Chemical Education, 91*(8), 1191–1197.

Papert, S. (1980). *Mindstorms: Children, computers and powerful ideas.* New York, NY: Basic Books Inc.

Psycharis, S., & Kallia, M. (2017). The effects of computer programming on high school students' reasoning skills and mathematical self-efficacy and problem solving. *Instructional Science, 45*(5), 583–602.

Scherer, D., Dubois, P., & Sherwood, B. A. (2000, Sept./Oct.). Vpython: 3d interactive scientific graphics for students. *Computing in Science & Engineering, 2*(5), 56–62.

Scherer, P. (2010). *Computational physics: Simulation of classical and quantum systems.* Heidelberg & New York, NY: Springer.

Sherin, B. L., (2001). A comparison of programming languages and algebraic notation as expressive languages for physics. *International Journal of Computers for Mathematical Learning, 6*(1), 1–61.

Stake, J. E., & Mares, K. R. (2005). Evaluating the impact of science-enrichment programs on adolescents' science motivation and confidence: The splashdown effect. *Journal of Research in Science Teaching, 42,* 359–375.

Taub, R., Armoni, M., Bagno, E., & Ben-Ari, M. (2015). The effect of computer science on physics learning in a computational science environment. *Computers & Education, 87,* 10–23.

Van Merriënboer, J. J., & De Croock, M. B. (1992). Strategies for computer-based programming instruction: Program completion vs. program generation. *Journal of Educational Computing Research, 8*(3), 365–394.

Vygotsky, L. S. (1978). *Mind in society.* Cambridge, MA: Harvard University Press.

Weatherford. S. (2011). *Student use of physics to make sense of incomplete but functional VPython programs in a lab setting.* Doctoral Dissertation (unpublished).

Weintrop, D., Beheshti, E., Horn, M., Orton, K., Jona, K., Trouille, L., & Wilensky, U. (2016). Defining computational thinking for mathematics and science classrooms. *Journal of Science Education and Technology, 25*(1), 127–147.

Wertsch, J. V. (1984). The zone of proximal development: Some conceptual issues. *New Directions for Child and Adolescent Development, 1984*(23), 7–18.

Wilensky, U. (1999). NetLogo. [Computer software]. Evanston, IL: Center for Connected Learning and Computer-Based Modeling, Northwestern University.

Wilensky, U. (2003). Statistical mechanics for secondary school: The GasLab multi-agent modeling toolkit. *International Journal of Computers for Mathematical Learning, 8*(1), 1–41.

Wilensky, U., & Reisman, K. (2006). Thinking like a wolf, a sheep, or a firefly: Learning biology through constructing and testing computational theories—An embodied modeling approach. *Cognition and Instruction, 24*(2), 171–209.

Winslow, L. E. (1996). Programming pedagogy—A psychological overview. *SIGCSE Bulletin, 28*, 17–22.

Xiang, L., & Passmore, C. (2015). A framework for model-based inquiry through agent-based programming. *Journal of Science Education and Technology, 24*(2–3), 311–329.

CHAPTER 6

PERSONIFICATION OF PARTICLES IN MIDDLE SCHOOL STUDENTS' EXPLANATIONS OF GAS PRESSURE

Robert C. Wallon and David E. Brown

Students often attribute human characteristics to non-human entities when reasoning about scientific phenomena. This process is called personification, and the traditional view is that this is undesirable in science education because it indicates misconceptions. Although arguments have been raised in favor of allowing personification in science classrooms for purposes of equity, this chapter illustrates specific conceptual affordances of personification. We argue personification can function as a bridge to build and communicate knowledge about complex physics phenomena, such as gas pressure, which may be challenging early in the learning process due to barriers of scientific terminology or formalisms.

Keywords: analogies, conceptual metaphor, dynamically emergent conceptions, embodied cognition, particulate nature of matter, personification

Physics Teaching and Learning: Challenging the Paradigm, pages 135–151.

Now the drop of water extends about fifteen miles across, and if we look very closely at it we see a kind of teeming, something which no longer has a smooth appearance–it looks something like a crowd at a football game as seen from a very great distance.

—*Richard Feynman*

The quote opening this chapter comes from Richard Feynman's famous *Lectures on Physics* in a section where he communicates about the particulate nature of matter by urging the reader to imagine a magnified view of a drop of water (Feynman, Leighton, & Sands, 2011, pp. 4–5). He communicates the idea that particles are in constant motion by comparing them to a crowd of people at a football game. His example is illustrative, drawing on an experience the reader has likely had in order to develop the concept of particles. But his example also ascribes human qualities to inanimate objects, an offense traditionally viewed as out of bounds in the science classroom. This brief example helps set the stage for the central aim of this chapter in which we explore students' use of personification in physics learning. Personification occurs when human qualities are attributed to non-human entities. For the purposes of this chapter, we use the term personification broadly to include several more nuanced concepts such as animism, anthropomorphism, and teleology. The common assumption is that personification is a barrier to science understanding. Consequently, teachers are advised to reduce or eliminate students' use of personification when it occurs in science classrooms. In this chapter we question this common assumption and explore how personification can provide conceptual resources helping students make sense of scientific phenomena.

REVIEW OF RESEARCH

We conceptualize personification as a specific instance of analogical thinking and metaphor, and therefore we briefly provide context for these ideas in science education and science. Then, we discuss literature explicitly addressing personification. Lastly, we comment on theoretical perspectives of embodied cognition and conceptions as dynamically emergent structures framing our work.

Analogical thinking involves explicitly thinking about one domain in terms of another domain by mapping features from a base to a target. Many researchers have argued this type of thinking has advantages in science education. For example, Duit (1991) argues analogies can promote conceptual change, provide ways to understand and visualize abstract ideas, and connect to students' prior knowledge. Indeed, studies have demonstrated instructional benefits of analogies used in various ways, including providing analogies to students (e.g., Paris & Glynn, 2004; Spiro, Feltovich, Coulson, & Anderson, 1989), asking students to create analogies to support their scientific explanations (e.g., Wong, 1993), and co-constructing analogies with students (e.g., Clement, 2013).

In addition to the intentional use of analogy in science instruction, researchers have argued metaphor subconsciously underpins human thought more broadly. For instance, Lakoff (1993) argues abstract concepts we encounter on a daily basis like time, change, causation, and purpose are metaphorical. In an extension of this viewpoint, some researchers have claimed that "thinking about and understanding science without metaphors and analogies is not possible" (Niebert, Marsch, & Treagust, 2012, p. 849).

It is important to note the importance of analogy is not restricted to the context of school science. There is evidence that analogical thinking is prevalent for practicing scientists. For example, Dunbar (1997) has shown scientists use analogies while trying to solve problems during laboratory meetings. Brown (2003) has argued metaphorical thinking is at the core of various scientific activities, such as when scientists "design experiments, make discoveries, formulate theories and models, and describe their results to others" (p. 14). And Clement (1988) found analogies to be pervasive in scientists' problem solving. Thus, analogical thinking and metaphor are important for thinking about and understanding science not only for students but also for practicing scientists.

Despite the potential affordances of analogies for science instruction and the argument that thinking without metaphor in science is impossible, this type of thinking has been criticized as a potential source of misconceptions. It has been suggested students may transfer a misconception in a base domain to a target domain or students may focus on less relevant surface features rather than more relevant deep features while thinking analogically (Duit, 1991).

Nowhere have criticisms of analogical thinking in science been as intense as in the context of personification, a type of analogical thinking involving attributing human characteristics to non-human entities. While personification, and the associated calls against it, may be most common in the domain of biology, it is also present in the physical sciences, as evidenced by Taber and Watts' (1996) exploration of students' personification as it relates to chemical bonding. To exemplify the criticisms of personification, Hughes argued against personification in a 1973 issue of the practitioner journal *Science and Children*, while at the same time admitting scientists may use personification out of convenience. Hughes claimed personification "is never needed for a complete understanding of the behavior of a nonhuman species" (p. 10). This claim holds that mechanistic causal explanations are incommensurate with personified explanations.

Yet, Brown (2003) gives several examples of claims from scientific papers rooted in personification. For instance, one example is, "These microbes infect cells and enlist several of the components that cells normally use to extend lamellipodia to power the bacteria's own travels within the host's cytoplasm" (p. 48). Brown argues this example shows bacteria performing actions in a purposeful manner, which is rooted in our understanding of the intentionality involved in many human actions. This example calls into question Hughes' claims that personification can be eliminated from scientists' thinking.

If scientists use personification in productive ways, then it may be worth exploring the potential affordances of using personification with students. Some researchers have advocated the use of personification in science classes for participatory purposes. Watts and Bentley (1994) argued the exclusion of personification in science classes could dehumanize the subject and consequently could limit the participation of females in science. Allowing personification in science classes for their instructional value was recommended by Zohar and Ginossar (1998), but only when accompanied by engaging students in explicit discussions about what is truly meant when speaking in such terms.

Rather than barring personification in the science classroom, it seems the most recent views are to exploit its value for instructional purposes while ensuring students recognize those instances as metaphorical. This use of personification has been labeled as *weak* to contrast with *strong* personification, where a student may "think a sodium atom literally experiences its world through feelings and emotions much like hers" (Taber & Watts, 1996, p. 564).

Thus far we have reviewed literature to present contradictions framing this study. The primary contradiction is that personification has arguments both for and against its use in the science classroom because of its potential to be positive and negative for science learning. A secondary contradiction is the existence of an ideal of science without personification, but examples from the authentic practice of science show personification abounds in the domain (e.g., Brown, 2003; Clement, 1988; Dunbar, 1997). Regardless of these contradictions, personification shows up as a common type of metaphorical thinking for students and scientists, thus warranting attention from the educational research community.

Much of the literature discussed in this review spans the last several decades. The reader may be wondering how the topic of personification connects to more recent work in educational research. We would like to revisit the topic of personification in light of growing interest in the theoretical position called embodied cognition, which claims human thought processes shape and are shaped by bodily experiences (Glenberg, 2010). From this theoretical perspective, there is value in considering the mutual influence of thought and body in students' science learning.

Students' bodily experiences may be brought to bear on making sense of scientific phenomena, which is particularly relevant when conceptions are viewed as dynamically emergent structures (Brown, 2014). From this perspective, conceptions emerge in a given context from aspects of knowledge that can vary in several ways, including their source. Brown (1993) describes conceptions at various levels, including the level of core intuitions developed from patterns of agentive experience in the world. These perspectives inform our aim to take seriously instances when students use personification and to consider the functions of these instances in students' conceptual ecologies.

THE PRESENT STUDY

As members of a research team developing computer simulations to help students construct causal mechanistic explanations of gas pressure, we noticed students would often use personification when they spoke about their ideas concerning molecules. We became curious about their use of personification and wondered whether it was productive or counterproductive for their development of causal mechanistic models of gas pressure. This study is a subset of the larger project (which is elaborated upon in the next section). It is intended to examine the personified ideas middle school students use to explain molecular behavior of a gas. In the present study we specifically explore the following questions:

1. How do students use conceptual metaphors underlying instances of students' personification of molecules?
2. Can affordances and constraints of those conceptual metaphors on students' understanding of gas pressure phenomena be identified?

Method

The data for this study were from a larger data corpus collected during the first year of a larger project called GRASP (GestuRe Augmented Simulations for supporting exPlanations). The main goal for the first year of the GRASP project was to learn more about students' intuitive ways of explaining specific scientific phenomena, including verbal and nonverbal components. The phenomena of interest relate to gas pressure, heat transfer and seasons. While doing a systematic analysis on the quality of students' explanations for the broader project, we noticed several instances when students talked about molecules having human characteristics with regards to feelings and goal-directed actions. These instances were identified for gas pressure and selected for further qualitative analysis in the present study.

The primary data sources used in this study are video recordings and associated transcripts of interviews with individual middle school students as they shared their developing explanations of gas pressure. Interviews were conducted with 24 middle school students in various settings outside their normal class time such as resource periods, after school clubs, and laboratory settings. Student explanations were focused on explaining phenomena related to air in a sealed syringe (Figure 6.1). These phenomena included noticing the plunger of the syringe could be compressed, the plunger became harder to push the farther it was pressed, and once released, the plunger moved to its original position. Interviews typically lasted 20–30 minutes.

Episodes during the interviews when students used explicitly personified ways of talking about molecules were identified from transcripts and selected for further analysis. Four of these episodes were identified, and they are the focus of the four cases explored in the next section. These episodes were selected because the personified speech seemed to be at the core of student thinking rather than a poetically expressive way of talking. An interpretive case study approach (Creswell,

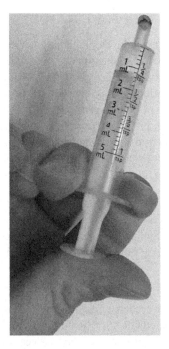

FIGURE 6.1.   Air Being Compressed in a Sealed Syringe

2013) was used to answer the research questions. Such an approach is appropriate because of the bounded focus on student thinking in a very specific way (i.e., with personification). When examples of personified speech were identified, excerpts of the video were analyzed to identify how students used personified thinking and the extent to which students' ideas were productive or counterproductive for their understanding of gas pressure. Personified thinking was considered productive when students were able to use it to identify molecular mechanisms that could account for the phenomena they observed, and accordingly, personified thinking was considered unproductive when it did not result in connections to molecular mechanisms. We found multiple examples of such thinking, which are presented in four cases in the following sections. In one of the cases, personification was conceptually unproductive, while in three cases personification was productive.

RESULTS

Case Studies

Case 1: Tracey Draws on Experiences Feeling Hot in Tight Spaces

Tracey is a student (indicated in the transcript excerpts with T, the first initial of her name) explaining to the interviewer (marked I in the transcript excerpts) what happens with the molecules when the plunger of the syringe is compressed.

T: They're just bouncing around like normal right now, and then when I do this [pushes plunger down] they get faster, and faster, and faster, and then I can't hold it any more so then it goes back to normal because they are getting like...they don't like small spaces and tight spaces, so they want to escape....yeah.

I: Why do you think they don't like the tight space?

T: Because...because they're very vigorous when they get in a tight space, and I guess that causes friction, and the heat doesn't want to...the heat molecules want to expand.

In the first excerpt Tracey brought up an idea we see expressed in an upcoming case by Blake; the idea that the molecules want to escape from tight spaces. When pressed on why she thinks molecules don't like tight spaces, her reasoning has to do with a non-canonical idea that the gas molecules get hot due to friction. Later in the interview she elaborates on her idea, and begins describing the molecules in personified terms.

T: It's closer when they get really fast, and then when they get... when they get away, then it's okay because the heat is not as much anymore. When you're in a tight space the heat...the...you get really hot easily.

From this excerpt we see Tracey's reasoning about the molecules heating up comes from her human experiences of feeling warm in a tight, crowded space. Although it can be productive to connect pressure and volume of a gas to the temperature of a gas in certain instances, such as thinking about the ideal gas law, the causal connections between molecular motion and temperature are not clear in this excerpt. Rather, Tracey's ideas are related to the ideas that the molecules are "feeling hot" when they are in tight spaces. Because she is unable to use these ideas to account for the causal explanation of why the plunger of the syringe moves back out when she lets go, this episode is considered to be unproductive for her thinking about the phenomena. Tracey's metaphor is represented in Table 6.1.

Tracey's case demonstrates what those who oppose students' use of personification are afraid of: her use of personification leads her down a conceptual path that differs from the intended learning outcomes. The next three cases, however,

TABLE 6.1. Mapping of Bases and Targets in Tracey's Personified Reasoning

Base	Target	How the Metaphor Influences Thinking About Gas Pressure	Evidence from Transcript
Humans getting warm in a tight space	Movement of gas molecules in a small volume	Promotes the non-canonical idea that gas molecules feel warm when they are in a tight space like how people feel warm in a tight space	"When you're in a tight space the heat...the...you get really hot easily."

illustrate instances when students' use of personification was helpful, even when it did not initially start that way.

Case 2: Amanda Connects to Walking in Crowded Hallways

Amanda has seen that a sealed syringe filled with air will compress and then will pop back out when the plunger of the syringe is released. The interviewer asks the student what she thinks is happening inside the syringe when it is compressed.

I: So, I'm curious, if, imagine that you had some magical glasses that let you look really, really close at the air that you said was in there. What do you think you'd see?

A: Well, you'd probably see the molecules, umm, moving faster because they're. Well, probably not moving as much because they're really compressed. And, they're like, so if they're moving like this, they're going about their business, then you put the air they're gonna like.

I: So when it's, when it's just sitting like this. Show me what you think.

A: They're probably just like moving around like this, letting the air move around.

I: Okay. And, then, when you push the plunger down.

A: Then they're probably in, like, packed together and start, like, shaking like that. And, then it's gonna all compress. And, they're gonna wanna, like, get apart and have their own room, so you let go and it's gonna, like, they're gonna let go, pretty much.

I: So, why is it, umm, you said that, umm they're gonna want their own room. Why is it that the molecules would want their own room?

A: Because they can't move as much. Like I know if I'm packed in a tight space. Like when I'm walking through my sixth grade hallway, all these people are in there, walking around talking. I'm like no. I need to get to my class. And so, I like to put my arm out and say move. Whoever runs into this hand is gonna get run over.

Amanda brought up her own experiences trying to walk through crowded hallways to help her reason about what molecules of gas do when they are in a smaller volume. Using her experiences led her to the non-canonical view of less molecular motion in a smaller volume. She also, however, specified in her example, that she would be walking with a hand outstretched and that "whoever runs into this hand is gonna get run over." This example indicated she would collide with other people, although it would be unintentional. The example seems to be a way for Amanda to make sense of the more frequent yet random collisions between molecules when in a smaller volume. These metaphors are represented in Table 6.2. The interview continues with the interviewer asking Amanda to think about walking through a relatively empty hallway and to compare that experience to the situation with the syringe.

I: And, what if it's not a passing period? What if you get a pass from your class? What's it like going through the hallway?

A: It's usually like, like really nice and easy because there's not that many people sitting out in the hallway. People don't get in trouble that much, and they can't send them out there anymore. But like, it's, I can just, like, walk straight through pretty much. And, only see like three, four people through my hallway. And then maybe five people in the seventh grade hallway. It's just, like, real nice and open and I like that. Like I had to move from my seat today cause there was too many people around me. Like get away from me.

I: And, what's that like your examples with either the crowded or non-crowded hallways? When it's like this [shows syringe with plunger pushed out]?

A: When it's like that then, in my hallway it's like me getting a pass just to, like, walk down the hallway.

I: And, when we push the plunger in?

TABLE 6.2. Mapping of Bases and Targets in Amanda's Personified Reasoning

Base	Target	How the Metaphor Influences Thinking About Gas Pressure	Evidence from Transcript
Humans in a crowded space	Gas molecules in a small volume	Promotes the non-canonical idea that gas molecules move less in a small volume like people in a crowded room	"Because they can't move as much. Like I know if I'm packed in a tight space."
Walking through a crowded hallway with a hand outstretched	Randomness of molecular collisions	Promotes the canonical idea that gas molecules collide with other molecules in random, non-purposeful ways	"And so, I like put my arm out and say move. Whoever runs into this hand is gonna get run over."
Walking through a less crowded hallway	Frequency of molecular collisions with lower relative density of molecules	Promotes the canonical idea that gas molecules collide less often when they are less densely packed	"But like, it's, I can just, like, walk straight through pretty much."

A: That's, like, me being crowded, putting my arm out, trying to move, get somewhere in the hallway.

In the second excerpt from the interview we saw Amanda was able to map her hallway example to the volume of the syringe. The interviewer used her metaphor of walking through a crowded hallway to help co-construct a metaphor of walking through a less crowded hallway. Her example helped her think about the relative abundance of molecules per unit area when the syringe is uncompressed and compressed by thinking about there being less students in a clear hallway and more students in a crowded hallway. Thinking about the gas molecules in this way helped Amanda consider how often collisions between molecules would happen when they were packed more densely or less densely. At the end of the interview in an episode not reflected in the presented transcript, Amanda is able to connect this idea to molecules colliding with the container as part of her explanation. This metaphor is also represented in Table 6.2.

Case 3: Blake Calls Molecules Claustrophobic

Blake also is discussing the syringe gas pressure phenomena with the interviewer.

I: Okay. So it sounds like as, as it gets smaller and smaller space the particles are...

B: Like become expanded, but it's, but they're not like multiplying. It's just. It looks like they're getting bigger and bigger, which is stopping it. But they're not. Because they're moving.

I: So. I think, I think maybe when you say expanding you mean maybe they're pushing out more? Is that what they're trying to do? How are they pushing out more? What is, what is making the pushing?

B: Uh, cause how many there are together. It's how many there are together. It's, uh, it's harder to fit in that one size. So it's pushing each other like—we attend school—kids everywhere. And, it's pushing, people hitting and hitting.

I: When you're getting in a tighter space?

B: Yeah. And like if one kid's trying to get through, he won't be able to. And, he won't be able to push all of them. And, so he's gonna have to stay there. He's like, 'Well, I'm late for class.'

At this point in the interview Blake seemed to be drawing on an analogy similar to the analogy Amanda drew upon during her interview. He thought about molecules in terms of students moving through a school. Thus far, the metaphor seems to be counterproductive because it promotes the non-canonical idea that gas molecules move less in a smaller volume. Blake's metaphor is represented in Table 6.3. During the next part of the interview, Blake continues to talk about molecules in terms of students, such as running around at recess when the volume of the syringe is larger. In the next excerpt Blake is looking at a simulation showing gas molecules when the volume of the container they are in is decreased. As he is trying to make sense of what the molecules are doing, he says they are trying to escape.

B: Like. The way it tells me is like, though in a close space, they don't wanna be that close together. But, this is blocking it, which is preventing them. Which, they're claustrophobic.

I: But, tell me based on what you can see.

B: They're moving around, way too much. They're speeding up. They're trying to find ways to escape, but, they won't be able to.

I: How do you know they're trying to find ways to escape? What are they doing?

B: They're bouncing off the walls.

I: They're bouncing off walls. So, they're bouncing off walls. And, how much bouncing off walls is happening now? [Volume is increased on the container in the simulation].

B: Not a lot. Not as much as it was. Yeah, there's bouncing off walls, but not as much. Like, see, that one, like they're just moving. [Volume is decreased on the container in the simulation]. Now it's bouncing off the walls quite a bit. And, they're bouncing off walls like crazy.

TABLE 6.3.   Mapping of Bases and Targets in Blake's Personified Reasoning

Base	Target	How the Metaphor Influences Thinking About Gas Pressure	Evidence from Transcript
Humans in a crowded space	Gas molecules in a small volume	Promotes the non-canonical idea that gas molecules move less in a small volume like people in a crowded room	"We attend school. Kids everywhere. And it's pushing the people hitting and hitting;" "And like if one kid's trying to get through it, it won't be able to. And it won't be able to push all of them. And so they're gonna have to stay there."
Claustrophobic humans in different sized rooms	Frequency of molecular collisions in different volumes	Promotes the canonical idea that gas molecules collide more often with the walls of their container if the volume of the container is decreased	"[Volume is decreased]. Now it's bouncing off the walls quite a bit. And now it's, it's ten. And they're bouncing off walls like crazy."

I: So what do you think this [points to image of pressure gauge] is measuring?

B: About how many times it bounces off the wall.

I: Yeah. I think you might be right. Right. So, what is air, what is air pressure?

B: About how many times it bounces off that wall. How many times the air bounces off the wall.

In this excerpt, Blake continued to think about molecules in personified terms. Interestingly, he said the molecules had claustrophobia, a fear of small places. Following through on that line of thought, he no longer expressed ideas about the molecules moving less in a smaller volume as we saw with the first excerpt presented from his interview. Instead, he described the molecules as frantically trying to escape when the volume is decreased. This idea seemed to be productive for him because it helped him notice the gas molecules collided more often with the walls of their container when the volume was smaller. Furthermore, he used that idea to eventually provide a definition of air pressure fairly aligned with a canonical view when he says that air pressure is "how many times the air [molecules] bounce off the wall." This metaphorical mapping is represented in Table 6.3.

Case 4: Lisa Imagines Blind People in a Room

Lisa is discussing what is happening with the molecules in the high volume and low volume scenarios.

L: Well, if it's very low like that one is, they have almost no space to move without running into something else, but, if it's increased a lot, there's a bunch of, like the volume is increasing so they have more freedom. Like, if there was...if you had a big old room, and you put a bunch of blind people in there. The bigger the room, the less likely they are to bump into each other because there's more, there are more options of places to be. Whereas, if it was a smaller place that you put people in, they would; they have less paths to take, and so they would be more likely to choose the same paths or paths that intersect.

I: So, what I'm hearing you say is that the number of times that the particles are bumping into each other is different, and when it's in a smaller space...

L: It's higher. They bump into each other more, and the bigger the space they bump into each other less.

Through her analogy of blind people moving around in different sized rooms, Lisa reasoned about the relative motion of molecules in different volumes. It's interesting to ask why she stipulated blind people specifically in her analogy. Since

TABLE 6.4. Mapping of Bases and Targets in Lisa's Personified Reasoning

Base	Target	How the Metaphor Influences Thinking About Gas Pressure	Evidence from transcript
Rooms of varying sizes with blind people walking around	Gas molecules in volumes of varying sizes	Promotes the canonical idea that gas molecules collide more often in a smaller volume and that the collisions are random	"[When it's a smaller space] They bump into each other more, and the bigger the space they bump into each other less."
			."..so they would be more likely to choose the same paths or paths that intersect."

she talked about people bumping into each other, she seems to have realized that sighted people would see each other and intentionally avoid collisions. As such, she seemed to be actively trying to find a way to bridge from the intentional inter-actions of people milling about in a room to the random interactions of molecules moving around in a container. This metaphor is represented in Table 6.4.

Cross-Case Results

This section explicitly addresses each research question in light of all of the cases.

1. How do Students Use Conceptual Metaphors Underlying Instances of Students' Personification of Molecules?

A similarity in three of the four cases is that students seemed to be broadly using the conceptual metaphor *molecular motion is human motion*. Students, however, also used more specific metaphors nested within this broader category. One example of a more specific metaphor is molecular motion in a smaller vol-ume is human motion in a more crowded space (from Amanda and Blake). Other examples of more specific metaphors nested within the base of *human motion* included a) human motion with a hand outstretched (from Amanda), b) motion of claustrophobic humans (from Blake), and c) motion of blind humans (from Lisa). While Amanda, Blake, and Lisa focused on how humans *moved* in their personi-fied metaphors, Tracey focused on how humans felt in her personified metaphor. Rather than *molecular motion is human motion*, Tracey seemed to be using a metaphor of *molecular motion is human experience*, as she focused on how a person may feel uncomfortably warm in a tight space.

2. Can Affordances and Constraints of Those Conceptual Metaphors on Students' Understanding of Gas Pressure Phenomena be Identified?

Three of the students (Amanda, Blake, and Lisa) made significant progress in their understanding of the gas pressure phenomena through their use of personified thinking. One student (Tracey) did not make significant progress by using personified thinking. Initial attempts to answer this research question can contrast the case of Tracey with the cases of the other students. An apparent contrast in these cases was the differential use of a broad personified metaphor for framing their thinking about molecular motion. Amanda, Blake, and Lisa used a base of human motion. While this base was sometimes counterproductive (see further discussion below), congruency between the base and target enabled the students to further specify their metaphors in ways that became productive. Tracey used a base of human experience. It seems a base of human emotion alone was incongruent with a target of molecular motion. This case brings up a conjecture for further exploration: personified metaphors can perhaps be counterproductive for students when the bases do not easily allow for congruent mappings onto the targets.

It is worth noting Amanda's and Blake's initial uses of the personified metaphor that *molecular motion in a smaller volume is human motion in a more crowded space* was counterproductive. They both mapped the idea that humans have a more difficult time moving around in a densely crowded space to generate the non-canonical idea that molecules would not move as much in a smaller volume when they are spaced more densely. Simply having the congruency of human motion and molecular motion did not necessarily result in a productive metaphor. Left to their initial ideas, Amanda and Blake could have had the misconception that gas molecules in a smaller volume acted more like the particles of a solid.

Through further specifying their metaphors, Amanda and Blake came to productive uses of personification. Blake ascribed claustrophobia to the molecules. If simply left at this, his metaphor would have been more aligned with Tracey's connection to human emotion. However, Blake reasoned about the effects of claustrophobic molecules by stating they would collide more often with the walls of their container, and he was able to integrate this idea into his understanding of gas pressure.

Amanda and Lisa further specified their metaphors to focus on the randomness of molecular motion. Amanda specified by thinking about a human walking with an outstretched hand such that anyone who bumped into it would happen unintentionally. Lisa specified by considering the motion of hypothetical blind people and discussing the likelihood of them bumping into each other in rooms of various sizes. These further specifications of the *molecular motion is human motion* metaphor seemed to enable students to overcome the aspect of human motion as intentional motion and instead consider human motion in circumstances resulting in random collisions.

CONNECTIONS TO PRACTICE, IMPLICATIONS, AND HYPOTHESES GENERATED

In each of the four cases, students used personified analogies to help with sense making of the physical situation. Possible implications of this work for teachers and instructional designers are that we should temper the inclination to view students' personified thinking as wrong. Instead, it may be useful to seek to use students' naturally personified analogies to help them bridge existing, sometimes tacit, knowledge to the desired scientific knowledge. Perhaps in some instances teachers may even consider encouraging students to think with personification to bridge intuitive, embodied ideas to a salient feature of a desired mechanism. Doing so grants respect to the vast knowledge students bring to the classroom and may serve to effectively ground canonical scientific understandings in embodied intuitions.

While the findings from this study show personified thinking could result in misconceptions and non-canonical ideas among students in some instances, the findings also showed personified thinking was a component of the productive development of canonical understandings of gas pressure. In the case of Lisa, personified thinking enabled her to do a thought experiment in which she considered the random motion of molecules. In the cases of Amanda and Blake, personified thinking served as a bridge for them as they thought about molecular motion in terms of human motion that initially resulted in non-canonical understandings. Refining their ideas about the contexts for human motion helped them bridge their understanding to the randomness of molecular motion and collisions of molecules with the walls of their containers, respectively.

It is also interesting to note what prompted students to use personified thinking during the interviews. In most cases, students used personified thinking to respond to difficult follow up questions from the interviewer (e.g., "Why is it that the molecules would want their own room?," "How are they pushing out more? What is, what is making the pushing?," and "Why do you think they don't like the tight space?"). Therefore, personified thinking provided students with ways of making sense of complex topics.

These cases showed that whether or not a teacher intends to engage students in thinking with personification, students may draw on it as a resource for their own sense making. As a consequence of this expectation, a question for further investigation is to ask how to use personification most productively. Further exploration is needed to determine whether or not teachers may find it productive to attend to students' use of personification instead of rejecting it outright. Some initial suggestions for practitioners are: 1) when exploring students' use of personification, teachers should consider seeking to explore the congruency between the bases students are using to make sense of a target and 2) when incongruent, teachers should consider supporting students either by helping them identify other bases better aligned to the target or by helping them further specify the bases so they map on to relevant features of the target.

In some cases, like Amanda's, even when a student initially incorrectly mapped the personification to the phenomena, she was able to realign her mapping in a conceptually productive way. For Amanda, realignment happened when the interviewer asked the rationale for her personified speech: why would molecules want their own room? This instance illustrates a potential prompt teachers can use when they hear students use personified speech. By pressing for a rationale, teachers may support students in mapping relevant parts of their examples to the target concepts. In Tracey's case, the prompt, "Why do you think [the molecules] don't like the tight space?" was not sufficient to result in conceptual progress. This finding raises the question of whether the interviewer could have done anything to re-orient the metaphor to a more apt usage. Perhaps the interviewer could have asked Tracey to focus on movement, knowing that considering those ideas has been productive for other students. Additional research would be needed to address these questions more conclusively. It may very well be the case that different topics have some metaphors more conducive for productive mappings than others. Future studies should further explore this topic, along with classroom implementations.

ACKNOWLEDGMENTS

This material is based upon work supported by the National Science Foundation under Grant No. DUE-1432424. Any opinions, findings, and conclusions or recommendations expressed in this material are those of the author(s) and do not necessarily reflect the views of the National Science Foundation.

REFERENCES

Brown, D. E. (1993). Refocusing core intuitions: A concretizing role for analogy in conceptual change. *Journal of Research in Science Teaching, 30*, 1273–1290.

Brown, D. E. (2014). Students' conceptions as dynamically emergent structures. *Science & Education, 23*, 1463–1483.

Brown, T. L. (2003). *Making truth: Metaphor in science.* Chicago, IL: University of Illinois Press.

Clement, J. (1988). Observed methods for generating analogies in scientific problem solving. *Cognitive Science, 12*, 563–586.

Clement, J. (2013). Roles for explanatory models and analogies in conceptual change. In S. Vosniadou (Ed.), *International handbook of research on conceptual change* (2nd ed., pp. 412–446). New York, NY: Routledge.

Creswell, J. W. (2013). *Qualitative inquiry and research design: Choosing among five approaches* (3rd ed.). Thousand Oaks, CA: SAGE.

Duit, R. (1991). On the role of analogies and metaphors in learning science. *Science Education, 75*, 649–672.

Dunbar, K. (1997). How scientists think: On-line creativity and conceptual change in science. In T. B. Ward, S. M. Smith, & J. Vaid (Eds.), *Conceptual structures and processes: Emergence, discovery, and change* (pp. 461–493). Washington D.C: American Psychological Association Press.

Feynman, R. P., Leighton, R. B., & Sands, M. (2011). *Six easy pieces: Essentials of physics explained by its most brilliant teacher.* New York, NY: Basic Books.

Glenberg, A. M. (2010). Embodiment as a unifying perspective for psychology. *Wiley Interdisciplinary Reviews: Cognitive Science, 1,* 586–596.

Hughes, A. (1973). Anthropomorphism, teleology, animism, and personification—Why they should be avoided. *Science and Children, 10,* 10–11.

Lakoff, G. (1993). The contemporary theory of metaphor. In A. Ortony (Ed.), *Metaphor and thought,* (2nd ed., pp. 202–251). New York, NY: Cambridge University Press.

Niebert, K., Marsch, S., & Treagust, D. F. (2012). Understanding needs embodiment: A theory-guided reanalysis of the role of metaphors and analogies in understanding science. *Science Education, 96,* 849–877.

Paris, N. A., & Glynn, S. M. (2004). Elaborate analogies in science text: Tools for enhancing preservice teachers' knowledge and attitudes. *Contemporary Educational Psychology, 29,* 230–247.

Spiro, R. J., Feltovich, P. J., Coulson, R. L., & Anderson, D. K. (1989). Multiple analogies for complex concepts: Antidotes for analogy-induced misconceptions in advanced knowledge acquisition. In S. Vosniadou & A. Ortony (Eds.), *Similarity and analogical reasoning* (pp. 498–531). New York, NY: Cambridge University Press.

Taber, K. S., & Watts, M. (1996). The secret life of the chemical bond: Students' anthropomorphic and animistic references to bonding. *International Journal of Science Education, 18,* 557–568.

Watts, M., & Bentley, D. (1994). Humanizing and feminizing school science: Reviving anthropomorphic and animistic thinking in constructivist science education. *International Journal of Science Education, 16,* 83–97.

Wong, E. D. (1993). Self-generated analogies as a tool for constructing and evaluating explanations of scientific phenomena. *Journal of Research in Science Teaching, 30,* 367–380.

Zohar, A., & Ginossar, S. (1998). Lifting the taboo regarding teleology and anthropomorphism in biology education–heretical suggestions. *Science Education, 82,* 679–697.

CHAPTER 7

COMPROMISED PHYSICS TEACHING

Assessment Driven Teaching

Isaac Buabeng, Lindsey Conner, and David Winter

In this study, four New Zealand (NZ) physics teachers' conceptions about teaching are compared in relation to their physics' teaching practices. Case studies have been used to present the analysis of themes as well as a platform to discuss the system enablers and constraints that influenced their practices in action. The conceptions, beliefs, and/or views held by four exemplary physics teachers have been identified and compared, and more importantly, the contextual factors have been identified that support or constrain teachers enacting their beliefs. The main constraint was the current content-driven assessment for Senior High School physics. As well, the teachers compromised their beliefs about teaching due to additional contextual constraints. Lack of time to prepare interactive activities and the emphasis on knowledge recall in external high stakes assessments rendered their desire for "minds on" approaches that included building on the key competencies of the curriculum practically impossible to implement. It is proposed that until assessment changes from a heavy emphasis on conceptual understanding to assessing inquiry and problem-solving skills, teachers will continue to focus on what is assessed and rightly so, align their teaching approaches to what is valued by the system.

Physics Teaching and Learning: Challenging the Paradigm, pages 153–174.

Keywords: Assessment standards; conceptions about teaching; physics teachers; teaching practice; students' performance; assessment-driven teaching

This study was designed to explore conceptions and/or views about teaching held by New Zealand (NZ) physics teachers in a context of delivering the New Zealand Curriculum (NZC). The curriculum specifies generic drivers of instructional design, such as students being at the centre of the learning design process, that builds on research in learning sciences, such as information processing, situated cognition and self-regulated learning as these contribute to constructivist approaches to teaching and learning, as described by Conner (2013).

The NZC identifies key ideas students should encounter in their science education in order to understand, enjoy and interact with our natural, physical world and the wider universe (Ministry of Education, 2007). In the science learning area, students are expected to explore both how the natural physical world and science itself work so they can participate as "critical, informed and responsible citizens in a society in which science plays a significant role" (Ministry of Education, 2007, p. 17).

The NZC emphasizes competencies which highlight the importance of creating and encouraging reflective thought and action; enhancing relevance; facilitating shared learning; making connections to prior learning and experience; providing sufficient opportunities to learn and inquire into the teaching and learning relationship, which are key elements of inquiry-based learning (Ministry of Education, 2007, pp 34–35). Teachers are required to use the NZC, together with the NZ Qualifications Framework, to design their own localized learning programs for their classes and schools, to meet the needs of their students and communities (Education Review Office, 2012; Ministry of Education). Teachers are further, required to encourage "reflective thought and action" and "facilitate shared learning" (p. 34). Teachers need to conceptualize knowledge as "actively constructed by and not given to" students (Pillay, 2002, p. 14), with students' needs being identified and accommodated in the learning design as described by Conner and Sliwka (2014).

To make the objectives of the NZC achievable, schools are encouraged to keep assessment to levels manageable and reasonable for both students and teachers. Assessment Standards are prescribed by the NZ Qualifications Authority which provides a range of standards covering a multitude of physics content areas from which teachers can choose to offer their students, as assessment targets. Some standards are assessed by teachers as their physics course is delivered ("internally assessed standards") and others are assessed by national examination at the end of the school year ("externally assessed standards"). The NZC stresses, "not all aspects of the curriculum need to be formally assessed, and excessive high stakes assessment in Years 11–13 is to be avoided" (Ministry of Education, 2007, p. 41). Harlen (2010) accentuated high stakes assessment (such as national external standards examinations) result in what is taught being determined by what is assessed rather than by what is of value in adding to a growing understanding of key ideas and development of reasoning skills and attitudes.

Our study explored, qualitatively, conceptions about teaching held by four NZ high school physics teachers. These teachers' conceptions were examined in light of constructivist epistemology and the enablers and constraints mitigating against teachers delivering their ideal program for learning. The alignment between the physics teachers' conceptions about teaching and how people learn and their teaching practice was considered. The study was guided by the following three research questions:

1.   What conceptions about teaching were held by the four high school physics teachers participating in our study?
2.   How did these conceptions influence their teaching practice?
3.   What enabled or constrained teachers from enacting their beliefs?

BELIEFS AND CONCEPTIONS OF PHYSICS TEACHERS ABOUT PHYSICS

Various labels have been used to refer to teachers' conceptions about teaching, such as, views, beliefs, practical personal theory, orientation, and cognitive structures (Buaraphan, 2007; Gess-Newsome, Southerland, Johnston, & Woodbury, 2003; Hewson & Kerby, 1993; Koballa, Glynn, & Upson, 2005; Tsai, 2002). Even though there seem to be subtle differences in the meaning of these labels, Tsai (2002) used them interchangeably. Here, we have used conceptions, beliefs and views interchangeably to describe participant teachers' understanding and experiences about teaching and how these are informed, according to the following definition:

The set of ideas, understandings, and interpretations of experience concerning the teacher and teaching, the nature and content of science, and the learners and learning that the teachers used in making decisions about teaching, both in planning and execution (Hewson & Kerby, 1993, p. 7).

Drawing from Pajares (1992) general research into teachers' beliefs and conceptions, Mulhall and Gunstone (2008, p. 439) noted "beliefs travel in disguise and often under alias—attitudes, values, opinions, perceptions, conceptions, implicit theories, explicit theories, and perspectives." Generally, what people know and believe influences their actions and informs the choices they make in their everyday lives. Beliefs also inform how teachers engage in and go about their classroom practices (Loucks-Horsley, Stiles, Mundry, Love, & Hewson, 2010). Teachers' conceptions about how science is developed may be related to their beliefs about how to teach science and how students learn science, including learning physics. For example, Gallagher (1991) described the views of the nature of science held by 25 secondary science teachers in Michigan, USA, as "unsettling" (p. 124). Classroom observations showed all the teachers emphasized science as

a body of knowledge. As a consequence, they spent more time developing terminology than building relationships between concepts, rarely engaging students in
experimental laboratory work.

This is in contrast to where teachers might embrace constructivism as a basis
for planning learning experiences (Tobin & Tippins, 1993) where teachers see
their role as helping students to connect new knowledge with existing knowledge
to generate enhanced knowledge, understanding and application of knowledge.
Such constructivist approaches to teaching and learning place students at the centre and require students to actively engage in activities that support their learning
progress. Cognitive engagement of students and consequent achievement tends
to improve when these approaches are purposively designed as part of learning
experiences (Conner, 2013).

Teachers build cognitive structures about teaching based on their prior
knowledge and experiences (Hashweh, 1996; Hewson & Kerby, 1993; Koballa
et al., 2005; Ladachart, 2011). Such cognitive structures can act as points of
reference for teachers' current teaching practice (Koballa et al.). Hewson and
Kerby noted teachers were likely to choose instructional approaches aligned
with their conceptions about teaching to enable them to achieve their teaching
goals. Even though such conceptions about teaching are often resistant to change
(Buaraphan, 2007), teachers may often compromise their ideal and aspirational
conceptions about teaching because of contextual constraints, causing them to
"hold working or back-up conceptions about teaching" (Ladachart, p. 177).

Further examples from the literature indicate this is not a new issue. For example, the interrelationship between science teachers' conceptions about teaching,
learning science and the nature of science was investigated by Tsai (2002). The
study's analysis of interviews conducted with 37 high school science (physics and
chemistry) teachers, showed most had traditional beliefs about the teaching and
learning of science. Namely, that science is best taught by transmitting knowledge
from teacher to students (e.g. transferring of knowledge, giving firm answers, providing clear definitions, giving accurate explanations and presenting the scientific
truths or facts). The way teachers taught was aligned with what they thought science included and, therefore, how it should be taught.

Using the Maryland Physics Expectations (MPEX) Survey (developed by
Redish, Saul & Steinberg, 1998) and Reformed Teaching Observation Protocol
(RTOP) (Piburn et al., 2000), Mistades (2006) investigated beliefs about physics
teaching held by three physics teachers who were faculty members of De La Salle
University in the Philippines, seeking to determine how many of these beliefs
translated into classroom strategies and practices. It was found that teachers' beliefs influenced their actions and practices directly in terms of classroom practice.
The physics teachers who participated in the study viewed learning physics as primarily understanding underlying ideas and concepts rather than simply focusing
on memorizing equations and formulae. The classroom observation data, obtained
using the RTOP, supported this view. Mistades indicated that the teachers' lessons

highlighted fundamental concepts by giving specific examples, showing relationships between concepts, and moving from simple to complex problems.

More recently, physics instruction in the south eastern USA was reported as being generally teacher-oriented, with lectures forming a significant part of the lessons (Sunal et al., 2015). In this case, classroom observation data (using the RTOP) did not support the teachers' indications during interviews where teachers indicated they thought hands-on-learning was important. They did not plan for, nor implement, hands-on-learning. Therefore their planning did not match their beliefs about learning.

As part of the research into teachers' knowledge and thought patterns (conceptions, beliefs and views) about teaching, some studies have drawn conclusions about teachers' practices that are unsupported by observational data (Hashweh, 1996; Tsai, 2002). Later work on teachers' conceptions and/or beliefs, however, including classroom observations, found a relationship between their conceptions about teaching and learning science, their epistemological beliefs and their teaching practice (Ladachart, 2011; Mulhall & Gunstone, 2008, 2012; Tsai, 2007). Therefore, in this study we deemed it important to consider whether teachers took action on their perceptions and beliefs or whether they were constrained in some ways to enact them.

The approaches used by physics teachers to teach physics were generally linked to their views about physics according to Mulhall and Gunstone (2008, 2012) who used qualitative methodology to explore views about physics held by a group of physics teachers whose teaching practice was traditional. These teachers' views were compared with the views held by physics teachers who used conceptual change approaches. The researchers found the perception that particular physics teaching approaches may be linked to particular views about physics "seemed to apply to the traditional group but not to the conceptual group" (Mulhall & Gunstone, 2008, p. 456). In discussing the implications of the data collected, the authors indicated persistent traditional approaches to teaching physics, which often failed to promote adequate student understanding of physics ideas. They posed a challenge to find ways of promoting teacher change, of helping physics teachers to understand and implement student-centered ways of teaching that may lead to better student outcomes.

There is no doubt that content knowledge is important and highly valued in physics assessments. When content knowledge is emphasized, there is a higher degree of focus on knowledge recall and the ability to reproduce content knowledge rather and on learning for understanding or application (Burnett & Proctor, 2002; Diseth, Pallesen, Hovland & Larsen, 2006). Given that teachers have their personal conceptions about teaching and these beliefs are likely to influence their instructional decision-making, we wanted to explore the conceptions held by some physics teachers and to examine them in the context of the constructivist epistemology enshrined in the NZC. We also wanted to explore the contextual factors constraining teachers from acting on their beliefs.

METHOD

Our work was conducted as part of a larger study (Buabeng, 2015) concerned with the lived experiences (Cohen, Manion, & Morrison, 2007) of physics teachers. A case study methodology was adopted. Case study approaches provide "many opportunities and strategies to gain insight into events that occur within the school and classroom" (Heitzmann, 2008, p. 523). The embedded multiple-case study design (Gray, 2009; Yin) was specifically adopted for our study because independent conclusions arising from two or more cases are more trustworthy than those from a single case (Yin, 2009).

Four exemplary physics teachers (three males and one female) from four secondary schools (two state schools, one integrated school and one independent school) in Christchurch, NZ, voluntarily participated in the study. They were a convenience sample geographically for on-going observation (Creswell, 2007). The teachers were identified and selected with the help of local science advisors. Reasons for selecting the schools included accessibility and willingness of school leaders and staff to participate. The physics teachers at these schools were interviewed and observed while teaching senior physics classes.

The research instruments for data collection were interviews and classroom observations. Semi-structured interview questions were designed to enable probing the teachers' views and opinions and this allowed the respondents to develop and expand on their own responses (Gray, 2009). The semi-structured nature of the interviews also allowed issues of particular concern to be discussed (Fraenkel, Wallen, & Hyun, 2012). Further questions, which were not expected at the commencement of the interview, could also be asked as new issues arose (Gray).

A Classroom Observational Guide (COG) was developed to measure physics classroom practices, including teacher preparedness in terms of both the content and pedagogy, among many others. The COG was adapted from the five scales of the Reformed Teaching Observation Protocol (RTOP) manual (Piburn et al., 2000). In this study, a 'non-participant observation' (Fraenkel et al., 2012; Gray, 2009) method was employed to investigate the subjects. In this type of observation, researchers are not directly involved in the situation being observed—instead they "sit on the side-lines and watch" without participating (Fraenkel et al., p. 446; Gray, 2009).

Eight observations were prearranged, however, some were missed due to interruptions and various school activities. This resulted in different numbers of observations being completed for each teacher: eight for Philip; eight for Nick; nine for Vicky and six for Bernard. With the exception of Philip, almost all the lessons observed lasted between 50–55 minutes. Philip had double periods for each class and so, these observations lasted 120 minutes. All the teachers were interviewed after the observations and all the interviews were audio recorded.

Data gathered during the interviews were transcribed verbatim and NVivo 10 for Windows (Lyn, 2012) was used to code the teachers' responses into nodes providing easy retrieval of the themes that emerged. The production of accurate and verbatim transcripts was integral to establishing the credibility and trustworthiness of our data. These were member checked with the participants for matters of accuracy and to provide an opportunity for the teachers to comment further.

Detailed descriptions were written during classroom observations as a record of what actually occurred. The individual case studies were prepared and using comparative analysis (Schwandt, 2001), the similarities and differences between the cases were identified. The comparison was helpful in determining how different contexts and individual expertise affected each teacher's Physics teaching.

Ethical approval for the study was obtained from our institution and all participants gave written informed consent.

RESULTS

In this section, the findings are presented from the case study teachers in terms of their conceptions about teaching, the teaching approaches they used, the constraining factors and their opinions about improving physics teaching and learning. We used pseudonyms to conceal their true identities. A summary of these detailed findings is presented in Appendix 1 for quick reference.

Conceptions about Teaching

The analysis of the classroom observations and Philip's interview indicated he held two main conceptions about teaching: promoting student engagement and establishing and maintaining good relationships with his students. These two conceptions originated from his initial teacher education program and had been consolidated through experience and reflection during his 30 years of teaching experience.

Promoting Student Engagement. According to Philip, the most effective learning occurs when students are highly engaged. As he stated:

> If they're engaged they're going to learn, and so, it's all about getting student engagement. If they become engaged, you can do practical work, you can do problem solving, you can do investigations, you can do discussion, you can do sometimes just an occasional lecture. (Philip)

During Philip's initial teacher education experiences, his course instructors favored student learning by inquiry and active participation in practical work. Philip valued this approach and believed it was a waste of time to have students copying copious notes mindlessly from the board (as he said). He believed that instead they should be active in doing things and thinking about what they were doing.

Establishing and Maintaining Good Relationships with Students. In addition to having students engaged in their work, Philip believed students were more motivated to learn if they have a good relationship with their teacher. He thought a good student-teacher relationship was vital because students would be more willing to ask questions to clarify lesson content they found difficult to understand. Philip's positive relationship with students was demonstrated through his demeanor in the classroom and how he valued students' opinions. He would sometimes sit by a student or group of students and provide individual and group feedback. In one of the lessons, Philip took four students, who were absent from the previous lesson, to the back of the room where they discussed the missed content while the other students completed individual work. He indicated he felt obliged to provide opportunities for them to catch up.

Nick held two main conceptions about teaching; sharing the historical narrative of physics with his students so they could see how concepts developed and providing learning opportunities for students to become more self-reliant and willing to engage in independent and collaborative learning. Nick claimed he developed the second conception at the time he was a teaching assistant at a university and that this conception was reinforced during his initial teacher education course.

Sharing the Historical Narrative of Physics. The history of science, and of physics in particular, had always been of interest to Nick. If he hadn't become a physics teacher, he stated he would have pursued a career as an historian. He described his belief that the easiest way to understand a scientific concept was to consider how it was first understood. He considered that the discoverer of something new had often had a valuable and simple way of trying to understand it, even if this understanding was later proved incorrect. Nick indicated telling his students about how discoveries and their interpretations were situated in a cultural and historical sense, was a good starting point for more challenging explanations of phenomena.

Providing Learning Opportunities for Students to Become More Self-Reliant and Willing to Collaborate. Nick said the best way to facilitate learning was to have students explain back what they understood to one of their peers. He recalled how he often heard the phrase "you truly don't understand something until you've taught it" during his own education and therefore applied this maxim to his teaching approaches. As he stated:

> To me the best indication of learning and the best learning that happens is actually when the students are helping themselves and helping others. I would very much like it if we had more time to do a lot more peer teaching. (Nick)

Even though Nick understood and advocated strongly for peer instruction, eight classroom observations showed that he rarely employed peer interaction during lessons. He explained it was quite difficult to implement active learning approaches, such as inquiry and experimentation, due to time constraints.

> There's not a lot of opportunity for inquiry and investigation. It's a time issue as much as anything because you have to give opportunity for them to learn themselves and then explain to somebody else, and you know, sometimes their explanations aren't good so you end up having to... (Nick)

On one occasion, Nick wanted his students to find out for themselves using guided inquiry how a capacitor functions. He recognized this would take much longer than the teaching time he had available so, he reluctantly implemented a transmissive approach to instruction.

Vicky held three dominant conceptions about teaching: students learn by doing, creating a supportive, inclusive learning community in the classroom and feeding students with content knowledge and detailed explanations. Her interviews revealed the first two conceptions had been established while she was acquiring her teaching qualification, whereas the third one had developed while teaching at her current school.

Students Learn by Doing. It was apparent Vicky valued hands-on activities and she wanted her students to engage in more of these in order to learn scientific knowledge through practical experiences. She indicated many times she would "like students to understand physics by doing experiments." "Learning by doing" was a pedagogical mantra promoted strongly in her initial teacher education program that Vicky had incorporated into her practice. She also valued collaborative learning and peer instruction to enhance students' understanding of concepts. Due to the emphasis on national summative assessments to measure student performance, however, she felt that opportunities for students to learn these ways was gradually being replaced by a transmissive, teacher-centered style of instruction. She recalled, "you learn by doing is a major philosophy to me but, I do this infrequently due to the pressure of the assessment and time constraints."

Learning Communities. According to Vicky, effective teaching and learning occurs when there is "an atmosphere of togetherness" in the classroom. Vicky reflected on her own unhappy experiences of learning physics at high school. She was determined not to subject her own students to this approach, as indicated by her comments:

> My physics teacher didn't try to connect with us in any way. There wasn't any sort of a nice atmosphere when you walked into the room. You walked into the room and he would be talking at you. There was no interaction. He just could not connect with us. I just really felt that there was no connection between the teacher and me. That, I mean I can sort of, it's hard for me to put myself back into my fifteen year old self, it just felt totally irrelevant in my life. I didn't do very well in physics because of him. I think I was, because I was always good at mathematics, that basically helped me pass. (Vicky)

Vicky was convinced that in order to help her students understand what they were learning, a good rapport between the teacher and the students was essential. Based on this conviction, she wanted to be accessible to her students so they would feel comfortable asking her even very basic questions. She explained it like this:

> …and setting up the environment that allows the kids to feel that if they don't get it they'll tell me, and in that way, having that close relationship. I know if things aren't working, because I can see it, and they're brave enough to tell me. Because it's okay to say in my classroom, I don't get it, let's try it again. Can you explain that in a different way because I still don't get it? (Vicky)

Vicky indicated this important aspect of teaching and learnng was missing during her own high school experience. In her role now as a teacher, she believed by including more relational aspects, she was more effective in promoting an atmosphere of cooperation.

Feeding Students With Content Knowledge and Detailed Explanation. It was Vicky's wish to have her students perform experiments to discover knowledge. She also wanted to be innovative in her practice to make physics as appealing as possible to her students. These aspirations were challenged by what she termed "the demands of the assessment system" and "students fixating on accumulating credits." She bemoaned the school and the students concern about the accumulation of credits to gain university admission rather than gaining a comprehensive understanding of physics concepts. Hence, rote learning was what her students preferred because they perceived it as the quickest way for them to pass their end of year examinations. She commented:

> They (students) expect me to tell them directly what they are supposed to know in order to pass the Year 12 and Year 13 exams. Because they're so used to learning to pass the standard. (Vicky)

Vicky's school focused heavily on students achieving academic success. Instructional activities not aligned with this goal were sometimes considered unnecessary or undesirable by the school administration. She described how she was very unhappy about this situation:

> …if you want to be a bit more innovative, it is terrifying…the battles I have to fight, to do this. And the fights I have to fight with my Head of Faculty, because her focus is getting the kids to pass the exams. The rote learning things, because that's what it says in the end of year exam. I'm horrified. And now I'm having these big fights with senior leadership. (Vicky)

The tension between Vicky's own pedagogical beliefs about effective practice and what she felt she was forced to enact in the classroom, was causing her to consider alternative employment options.

Bernard held two dominant conceptions about teaching; "seeing himself as a teacher of students rather than as a teacher of physics" and "helping students to become logical thinkers." Interviews with Bernard revealed the first conception was formed during his first few years of teaching and the second from participating in an in-service professional development program he attended soon after starting at his present school.

Seeing Himself as a Teacher of Students Rather Than a Teacher of Physics. According to Bernard, his job as a teacher was to help students reach their potential. The subject he was teaching was of secondary importance. He was happy to teach physics, chemistry, biology or mathematics as long as he could help students to reach their life goals. He claimed whenever he was asked "who are you and what do you teach?" his response had always been "I'm a secondary teacher, and I teach students." Seeing himself as a teacher of students had contributed to his teaching success because it had encouraged him to search for, and implement, pedagogical strategies that would ensure the success of his students across a variety of contexts.

Helping Students to Become Logical Thinkers. Bernard believed that effective learning includes the development of the skills and the ability to ask and solve deeper-level thinking-type questions, rather than just the superficial remembering of facts. He recognized that after his students completed their high school education, very few would use their physics content knowledge and so he wanted them to participate in activities that would promote critical thinking and develop the skills to continue independent thinking and learning long after leaving school. As he stated:

> I think giving students the ability to think deeply about things, ask questions, solve problems, I would define as an effective teacher or learner. The whole logical thinking skills and things is something we have to really push. (Bernard)

Bernard stated? that people with physics in their background, were often working in careers not directly related to science, such as banking and accounting and these people were often there because they were able to think logically in unfamiliar contexts. Bernard would have preferred his students to be engaged in activities that helped them develop as logical thinkers rather than lecturing to them and having them copying notes. Classroom observations confirmed Bernard's students were engaged in problem-solving and critical thinking activities at least weekly.

Teaching Practice

The conceptions Philip had about teaching clearly influenced his teaching approaches. Analysis of classroom observation data showed he used a variety of teaching methods to engage his students. He set up practical demonstrations and lessons for students. Sometimes, he lectured or provided detailed explanations and examples of physics problems on the white board. Occasionally, he played videos and used interactive demonstrations. He often linked physics concepts to real-world situations to make it more relevant for his students. A visit to a local playground helped to demonstrate the concept of angular momentum using a merry-go-round. When teaching the concepts of change in momentum and impulse, Philip took the students outside to the school field, where they were paired and given two eggs. The eggs were thrown from a distance by one student while the other student attempted to catch the egg with a cloth sheet. If the student was unable to catch the egg, it hit the ground and broke. When the lesson was reconvened indoors, Philip led a discussion about the demonstration and guided students in understanding the differences in the results with regards to the stopping force exerted by the sheet and the ground. The concepts were then discussed in the context of automobile accidents and the injuries sustained in relation to speed. The students seemed extremely excited and reflective of their learning during these episodes.

The observations of Nick's lessons about electricity and magnetism showed he often provided biographical information about well-known scientists to his students, with the aim of stimulating their curiosity by including the human interest side of physics. In one of his lessons on electromagnetism, Nick told the class about the life and contributions of Hans Christian Ørsted and Michael Faraday. Other stories included describing the work of the German physicist Gustav Kirchhoff and British engineer John Ambrose Fleming in separate lessons on Kirchhoff's circuit laws and the force on a current-carrying wire in a magnetic field respectively. Nick often included summary information on historical figures in the *PowerPoint* notes for his students

None of the lessons observed demonstrated Nick's second conception about teaching: providing learning opportunities for students to become more self-reliant and willing to help themselves and help others. He indicated a lack of time and the demands of the curriculum rendered this conception practically impossible to implement.

There was not much variety in Nick's teaching methods. His lessons were predominantly characterized by two episodes. First, there was a period of note-taking where students copied information from *PowerPoint* slides. While the students wrote, Nick explained the slides and sometimes provided further details and illustrations on the white board. In the second episode, students worked on exercises from textbooks or workbooks. At his interview, Nick stated that, due to limited time, there were still elements of rote learning in his teaching because there were basic facts and "things" students "just needed to know." He further stated students had to memorize or learn some "things" to be able to advance to the next step in their understanding of some topics and concepts which would be examined in the external standards assessments.

> Sometimes, it is just a case of knowledge transfer for sure… where they've just got to get this information, you know, you can't do as many of the good things as we might like to do. So, there are still elements of rote learning and chalk and talk, but there are some times when that seems to me to be the appropriate method. (Nick)

The nine classroom observations with Vicky showed a relationship existed between what she believed and her teaching practice. She took every opportunity available to have students perform projects and practical activities to support their learning. Students worked in groups of three or four on almost all these activities and sometimes they had the opportunity to plan their own investigations. When students undertook experiments, Vicky moved from one group to another, interacting with the groups and asking students "why" questions to stimulate their thinking. She wasn't just interested in questions and answers but rather encouraged students to reflect on the issues under investigation. She constantly asked students to explain their reasoning and elicited different responses from students to the same question, with discussions about the comparisons of answers.

As part of her quest to make physics more appealing, interactive demonstrations were sometimes inserted into lecture-type lessons. In one such lesson, Vicky described the operation of a generator and used a computer-based animation to support her explanation. Then, she encouraged the students to draw a labelled diagram showing all the forces involved. The students particularly enjoyed this approach. During her interview, Vicky described how she tried to show the connection between what was being taught and real-world physics applications.

Vicky encouraged and supported her students to become members of a learning community by setting up group work. She promoted positive interdependence and the negotiation of ideas. The use of group work happened most frequently during the practical parts of lessons, when students designed their experiments and the success of each group was valued.

Observations of six lessons on mechanics, and electricity and magnetism showed Bernard used a wide variety of teaching methods such as problem solving, simulations, demonstrations, discussions, and teacher-centered instruction, with the latter being the dominant teaching method.

Bernard strongly encouraged students to participate actively in thinking during his lessons. Rather than always introducing content via lectures, he organized exploratory activities for the students as a starting task. The "predict-observe-explain" strategy was used often to explore students' prior ideas. Activities promoting collaboration sometimes were embedded within tasks. When he was asked how he selected activities fitting within the limited teaching time he had available, Bernard explained his priority was to make physics relevant and interesting, while always keeping "an eye on the prize" which he described as being strong performance in examinations. He considered collaborative activities to be essential in supporting students to become logical thinkers. Bernard's conception of teaching as "helping students to become logical thinkers" was most visible in lessons incorporating practical work. Bernard considered the physics course as emphasizing historical aspects of physics knowledge far too strongly at the expense of covering newer and more interesting material. He also stated his physics course could be constructed with a much stronger mathematical emphasis but he tried to avoid this as it would make it too challenging for many students.

To support student learning, Bernard provided weekly investigations he called "Physics Olympics" where students competed to design experiments to investigate a physics problem and find a solution. Bernard spent a considerable amount of time discussing the technical aspects of performing these experiments and reviewing the relevant physics concepts. The tasks were carried out in groups of three or four students and active participation was encouraged. Each group of students had an opportunity to present its findings to the class. The group producing the correct or best answer was rewarded with a prize. During the post observation interview, Bernard described the Physics Olympics as a competitive and fun way to introduce physics concepts to students, especially those new to the subject.

Bernard perceived the amount of teaching time available for his subject to be a major constraint on his practice but, he believed he could still get students to both think deeply and achieve well in their examinations. He said he did not want to become a teacher who was in his words "dry in the delivery of the curriculum." He preferred to identify the content students needed to know and link that content to important life skills that could be taught through physics. As he said:

> I'd like to think that, I've got miles to go, trying to reflect an older subject like physics being taught in a more modern way and which is at least cognizant of some neuroscience and neuroscience research. (Bernard)

This statement indicated that he was more aware than the other teachers of the considerations for developing thinking skills to help students to learn more effectively.

Contextual Factors Supporting or Constraining Teachers Enacting their Beliefs

The general consensus among these four teachers was that assessment constituted the major factor constraining them to enact their beliefs about constructivism and deliver quality physics teaching in their respective high schools.

Vicky commented: "I think the biggest difficulty, the biggest roadblock by far, is assessment." She stated the focus on the importance of assessment was not only impacting on her as a teacher, but impacting on her students' attitude to learning because their priority had become the accumulation of credits to achieve the national qualification. Vicky described this situation as "very unfortunate."

Vicky and Nick both felt there was a conflict between the aspirations of The NZ Curriculum and the national assessment system. Nick viewed this conflict as a major hindrance to implementing his beliefs about teaching. He explained there was an expectation from most NZ universities that the content covered by key achievement standards would be taught in schools and that additional achievement standards would be included in a program of study to contribute to students' credits for entry into physics at universities. He described his physics program as "quite prescriptive," and as such, there was little or no time to explore content beyond the requirements for the assessment standards.

> There's my course, 23 credits (made up from multiple standards) which is quite prescriptive. I have to teach Kirchhoff's Laws, I have to teach internal resistance, I have to teach capacitors and inductors for DC, I have to teach capacitors and inductors for AC, I've got to do LCR circuits, I've got to do resonance. There's very little room and time to explore wider than that. (Nick)

In addition to the workload for designing and implementing the curriculum, Nick was frustrated by the rate at which his administrative workload was increasing and occupying time that could otherwise be used to prepare engaging physics lessons.

> The requirements for filling in forms even if they're electronic, is more than it was ten years ago and so we're ending up not spending as much time preparing interesting lessons. We're not spending as much time actually enjoying teaching in the classroom because we have so many other things on our plates. (Nick)

According to Bernard, there were a number of challenges associated with ensuring students experienced quality physics teaching and learning. The dominant challenges were the lack of alignment between the national curriculum and the achievement standards used for assessment, and the emphasis placed by the curriculum on traditional rather than modern physics content. Bernard was also very concerned about the current shortage of qualified physics teachers.

According to Philip and Bernard, internal assessment had placed additional workload on teachers and reduced the time available to them to spend on quality physics teaching and improving approaches to teaching physics. Philip believed that satisfying the administrative requirements for the national qualification had dominated professional learning for physics teachers for far too long.

Assessment was also the greatest constraining factor for Vicky, as she felt students were concerned with the accumulation of credits rather than learning. She viewed this situation as being in conflict with key aspirations of the national curriculum, which were to give schools the scope, flexibility and authority to design and shape their curriculum so learning and teaching was meaningful and beneficial to students. With the administrative requirements of assessment becoming more demanding, teachers had less time to properly prepare students for success in the assessments. These constraints, and others, made it practically impossible for physics teachers to incorporate the values and the key competencies of the national curriculum into lessons.

DISCUSSION AND IMPLICATIONS

This study not only investigated teachers' beliefs about teaching as previous studies have done (e.g. Hashweh, 1996; Tsai, 2002), but also used classroom observations to gauge whether the teachers acted on their beliefs. Other studies about teachers' beliefs that included classroom observations have found a relationship between beliefs about teaching and learning science and teachers' epistemological beliefs and their teaching practice (Ladachart, 2011; Mulhall & Gunstone, 2012; Tsai, 2007). In our study, the conceptions, beliefs, and/or views held by four exemplary physics teachers, were identified and compared, and more importantly, contextual factors were identified that enabled or constrained the teachers to enact their beliefs. It became evident that the conceptions about teaching held by the various participants in this study had developed over a number of years. The finding that teachers developed their conceptions over a longer period of time through multiple experiences, aligns with Ladachart's conclusion to this effect as well.

The relationship between the teachers' conceptions and their teaching practice in our study was observed in the actions they took. Nick's conception of sharing the history of physics to help students see how discoveries occurred and then developed, for example, he was observed to provide anecdotal connections with the history of physics and stories about scientists during his lessons. Bernard thought helping students to become logical thinkers was important. So, he spent more time than the other teachers in engaging students in activities where they found solutions to problems and developed critical thinking skills.

A study by Tsai (2007) of four teachers and their students that included classroom observations, found consistency between the teachers' conceptions about teaching and their teaching practice. Similarly, Mulhall and Gunstone's (2012) study of physics teachers found the teachers taught physics in a manner consistent with their views about teaching and learning. While there were examples of how teachers did put their beliefs into practice in this study, what was more surprising was that they often did not take action on their beliefs. In general, they wanted to align their practice with their understanding about effective teaching and learning, but often reverted to content delivery models to cover all of the content knowledge they assumed students needed to pass the assessment standards.

As indicated by Koballa et al. (2005), Ladachart (2011), and Tsai (2002), the context or conditions in which teachers teach, can have an influence on their conceptions about teaching and the extent to which these conceptions are enacted. These conceptions, argues Buaraphan (2007), about teaching are often resistant to change. Contextual constraints, especially assumptions about assessment requirements, may cause teachers to compromise or reinforce their conceptions about teaching (Buaraphan & Sung-Ong, 2009; Friedrichsen & Dana, 2005; Koballa et al.). As presented in our study, Nick and Vicky had a set of strongly-held ideas about teaching, including "providing learning opportunities for students to become more self-reliant and willing to engage in independent and collaborative learning" and "students learn by doing," respectively. Both teachers, however, compromised these beliefs. Lack of time and the demands of assessment rendered Nick's desire for minds on approaches practically impossible to implement. Vicky experienced the discordancy of her previously developed and held conception of students learning by doing, because of the emphasis in her school on success in external high stakes standards assessments. This discordancy led her to form a new conception about teaching: feeding students with content knowledge. She knew the aspirations of the curriculum were to integrate values and key competencies into student learning experiences, but she was frustrated that external pressures for her students to achieve acceptable results to gain the national qualification and admission to university, overrode an emphasis on competencies.

The findings in this study that Nick, Philip and Vicky compromised their ideal conceptions challenge the assertion by Buaraphan (2007) that conceptions are often resistant to change. Conceptions might be difficult to change, however we have presented examples of cases where systematic and local contextual factors work against the ideals teachers have for their work, so that they actively resist changing their approaches, even when they know and have evidence that more interactive and student-centered approaches might lead to more effective learning. The findings agree with the claim by Ladachart (2011) that contextual constraints may cause teachers to compromise their ideal and aspirational conceptions about teaching, leading to the formation of new and possibly less than ideal ones.

The NZC aspires to make learning relevant to learners, which the teachers in our study wanted to do. The participant teachers were frustrated because they felt that meeting the aspirations of the NZC had taken second place to satisfying the demands of the assessment system. Vicky was convinced that as long as students have to be assessed by examinations, the aspirations of the NZC were never going to be achieved in physics. As she said: "At the end of the day, I am being judged by how well they (students) do in tests, and they are being judged by how well they do in the test."

Evidence from this study shows the physics teaching of the four participants was driven by assessment. The assessment and its related administrative requirements required a lot of time from teachers they could otherwise have used to prepare quality lessons. The classroom observation data and the teachers' interviews suggested these teachers were always under pressure to complete assessment tasks. Therefore there was little or no time to personalize learning experiences for students, as aspired to in the NZC. The assessment standards for physics place a huge emphasis on recall of content knowledge, so this is what these teachers focused on.

IMPLICATIONS FOR POLICY AND PRACTICE

In order to shift pedagogical practices in physics teaching in NZ, there would need to be a change in what is emphasized in the national assessment standards. This change would require a shift from an overemphasis on content within assessments to assessing understanding, application, inquiry and problem-solving skills more than currently occurs. This change in approach would need to occur for physics to be valued by all stakeholders. It would be consistent with research showing systems change can support learners for their futures (Organization for Economic Cooperation and Development [OECD], 2013). If the assessment emphasis shifted, then teachers might be able to align the learning experiences they provided to students with their knowledge about how to help students learn using more constructivist, student-centered inquiry and problem-solving approaches. With better alignment between approaches to teaching and learning and assessment, teachers might be more likely to meet the outcomes promulgated in the curriculum document and help students to gain other key competencies simultaneously. Until assessment in physics is changed, it is highly likely that teachers will continue to align their teaching approaches to it and to what the assessment system for physics values.

APPENDIX 1: SUMMARY OF CASE STUDIES

Characteristics/ Case Names	Philip	Nick	Vicky	Bernard
Qualification	• Physics graduate and holds a Graduate Diploma.	• PhD in physics and holds a Graduate Diploma.	• Bachelor's degree in physics and a Graduate Diploma	• Master of Science (Marine Biology), Bachelor of Science and Graduate Diploma
Teaching Experience	30+ years.	12 years.	10 years.	25 years.
Conception about Teaching	• Promoting student engagement. • Establishing and maintaining good relationships with students.	• Sharing the historical narrative of physics to help students see how concepts developed. • Providing learning opportunities for students to become more self-reliant and willing to engage in independent and collaborative learning.	• Students learn by doing activities. • Creating a supportive, inclusive learning community. • Giving students content knowledge and detailed explanations.	• Seeing himself as a teacher of students rather than a teacher of physics. • Helping students become logical thinkers.
Teaching approaches	• Practical demonstration, • Problem solving, • Collaborative learning, • Simulations, • Lecture.	• Predominantly lecture with note-taking from PowerPoint, • Problem solving techniques.	• Hands-on activities, • Collaborative learning, • Interactive demonstrations, • Lecture.	• Problem solving, • Simulations, • Demonstration, • Discussion. • Collaborative learning, • Lecture.

Constraining factors (Decreasing order of importance as indicated by the teacher)	• Time constraints, • Assessment demands, • Alignment of achievement standards with the curriculum, • Increased workload, • Poor instruction in physics at junior levels, • Poor mathematical competency of students.	• Lack of time, • Dichotomy between curriculum and assessment, • Assessment demands and teacher work load, • Inadequately qualified physics teachers, • Poor public perception of physics, • Nature and structure of junior science.	• Premium on high stakes assessment, • Time constraints, • Assessments requirements, • Alignment of achievement standards with the curriculum, • Teacher workload, • Nature of the physics curriculum • Poor public perception about teaching profession	• Lack of qualified physics teachers, • Alignment of the curriculum to the achievement standards, • Students' preoccupation with assessments, • physics curriculum itself.
Way forward for improving upon physics teaching and learning	• Reduction in the number of assessments, • Reintroduction of a single internally assessed standard, • Improved remuneration and status of teachers, • Greater provision of professional development opportunities for teachers.	• Allocation of more time, • Re-alignment of achievement standards and curriculum, • Encouraging postgraduate physicists into teaching, • Greater provision of professional learning courses, • Better teacher remuneration.	• Reduction in assessment requirements, • Provision of better remuneration package for teachers, • Greater provision of professional development courses for teachers.	• Allocation of more time for teaching, • More qualified physics teachers, • Mentoring and teacher collaboration, • Provision of good support mechanisms for teachers and better remuneration.

REFERENCES

Buabeng, I. (2015). *Teaching and learning of physics in New Zealand high schools* (PhD Thesis). University of Canterbury, Christchurch.

Buaraphan, K. (2007). Relationships between fourth-year preservice physics teachers' conceptions of teaching and learning physics and their classroom practices during student teaching. *Songklanakarin Journal of Social Science and Humanities, 13*(4), 595–620.

Buaraphan, K., & Sung-Ong, S. (2009). *Thai pre-service science teachers' conceptions of the nature of science.* Paper presented at the Asia-Pacific Forum on Science Learning and Teaching.

Burnett, P. C., & Proctor, R. M. (2002). Elementary school students' learner self-concept, academic self-concepts and approaches to learning. *Educational Psychology in Practice, 18*, 325–333.

Cohen, L., Manion, L., & Morrison, K. (2007). *Research methods in education* (6th ed.). New York, NY: Routledge.

Conner, L. N. (2013). Students' use of evaluative constructivism: comparative degrees of intentional learning, *International Journal of Qualitative Studies in Education, 27*(4), 472–489, DOI: 10.1080/09518398.2013.771228

Conner, L. N., & Sliwka, A. (2014). Implications of research on effective learning environments for initial teacher education. *European Journal of Education, 49*(2), 165–177. [10.1111/ejed.12081] [Scopus]

Creswell, J. W. (2007). *Qualitative inquiry and research design: Choosing among five approaches* (2nd ed.). Thousand Oaks, CA: SAGE Publications, Inc

Diseth, A., Pallesen, S., Hovland, A., & Larsen, S. (2006). Course experience, approaches to learning and academic achievement. *Education and Training, 48*, 156–169.

Education Review Office. (2012). *The New Zealand curriculum principles: Foundations for curriculum decision-making.* Wellington: Education Review Office.

Fraenkel, J. R., Wallen, N. E., & Hyun, H. H. (2012). *How to design and evaluate research in education* (8th ed.). New York, NY: McGraw-Hill.

Friedrichsen, P. M., & Dana, T. M. (2005). Substantive-level theory of highly regarded secondary biology teachers' science teaching orientations. *Journal of Research in Science Teaching, 42*(2), 218–244.

Gallagher, J. J. (1991). Prospective and practicing secondary school science teachers' knowledge and beliefs about the philosophy of science. *Science Education, 75*(1), 121–133.

Gess-Newsome, J., Southerland, S. A., Johnston, A., & Woodbury, S. (2003). Educational reform, personal practical theories, and dissatisfaction: The anatomy of change in college science teaching. *American educational research journal, 40*(3), 731–767.

Gray, D. E. (2009). *Doing research in the real world* (2nd ed.). Thousand Oaks, CA: Sage Publications Inc.

Harlen, W. (Ed.). (2010). *Principles and big ideas of science education.* Gosport, Hants: Ashford Colour Press Ltd.

Hashweh, M. Z. (1996). Effects of science teachers' epistemological beliefs in teaching. *Journal of Research in Science Teaching, 33*(1), 47–63.

Heitzmann, R. (2008). Case study instruction in teacher education: Opportunity to develop students' critical thinking, school smarts and decision making. *Education, 128*(4), 523–542.

Hewson, P. W., & Kerby, H. W. (1993). *Conceptions of teaching science held by experienced high school science teachers.* Retrieved from http://files.eric.ed.gov/fulltext/ ED364426.pdf

Koballa, T. R., Glynn, S. M., & Upson, L. (2005). Conceptions of teaching science held by novice teachers in an alternative certification program. *Journal of Science Teacher Education, 16*(4), 287–308.

Ladachart, L. (2011). Thai physics teachers' conceptions about teaching. *Journal of Science and Mathematics Education in Southeast Asia, 34*(2), 174–202.

Loucks-Horsley, S., Stiles, K. E., Mundry, S., Love, N., & Hewson, P. W. (2010). *Designing professional development for teachers of science and mathematics.* Thousand Oaks, CA: Corwin.

Ministry of Education. (2007). *The New Zealand curriculum* Wellington: Learning Media Limited.

Mistades, V. (2006). *Linking teaching beliefs to classroom practice: A profile of three physics teachers.* Paper presented at the APERA Conference 2006 Hong Kong.

Mulhall, P., & Gunstone, R. (2008). Views about physics held by physics teachers with differing approaches to teaching physics. *Research in Science Education, 38*(4), 435–462.

Mulhall, P., & Gunstone, R. (2012). Views about learning physics held by physics teachers with differing approaches to teaching physics. *Journal of Science Teacher Education, 23*(5), 429–449. doi: 10.1007/s10972-012-9291-2

Organisation for Economic Co-operation and Development (OECD). (2013). *Innovative learning environments.* Paris: Educational Research and Innovation, OECD Publishing.

Pajares, M. F. (1992). Teachers' beliefs and educational research: Cleaning up a messy construct. *Review of educational research, 62*(3), 307–332.

Piburn, M., Sawada, D., Turley, J., Falconer, K., Benford, R., Bloom, I., & Judson, E. (2000). *Reformed teaching observation protocol (RTOP) reference manual.* Retrieved from http://www.public.asu.edu/~anton1/AssessArticles/Assessments/ Chemistry%20Assessments/RTOP%20Reference%20Manual.pdf

Pillay, H. (2002). *Teacher development for quality learning: The Thailand education reform project.* Bangkok: Office of the National Education Commission.

Redish, E. F., Saul, J. M., & Steinberg, R. N. (1998). Student expectations in introductory physics. *American Journal of Physics, 66*(3), 212–224.

Richards, L. (2012). *Using NVivo in qualitative research.* London, UK: Sage Publications.

Schwandt, T. A. (2001). *The Sage dictionary of qualitative inquiry*: Thousand Oaks, CA: Sage.

Sunal, W. D., Sunal, C. S., Dantzler, A. J., Turner, P. D., Harrell, J. W., Stephens, M., & Aggawal, M. (2015). *Teaching physics in our high school classroom.* Paper presented at the National Association for Research in Science Teaching (NARST) Annual International Conference, Chicago, IL, U.S.A.

Tobin, K., & Tippins, D. (1993). Constructivism as a referent for teaching and learning. In K. Tobin (Ed.), *The practice of constructivism in education* (pp. 3–21). Hillsdale, NJ: Lawrence-Erlbaum.

Tsai, C. C. (2002). Nested epistemologies: science teachers' beliefs of teaching, learning and science. *International Journal of Science Education, 24*(8), 771–783. doi: 10.1080/09500690110049132

Tsai, C. C. (2007). Teachers' scientific epistemological views: The coherence with instruction and students' views. *Science Education, 91*(2), 222–243.

Yin, R. K. (2009). *Case study research: Design and methods* (4th ed.). Los Angeles, CA: Sage Publications.

CHAPTER 8

COLLABORATIVE LEARNING WITH NETWORKED SIMULATIONS

Lisa Hardy and Tobin White

Modern computing and networking tools allow for new approaches to supporting collaborative learning in physics. While most computer-based simulations currently used in physics classrooms target individual or whole class instructional configurations, we focus on designing and investigating tools and activities to support learning in small groups. We present an approach to distributing simulations of physical phenomena across the devices of several students to support collaborative learning. Two empirical cases are presented to highlight some of the ways in which the networked and distributed nature of the designs influences student learning about physics concepts.

Keywords: physics education, collaborative learning, handheld computers, classroom networks, generative design, simulations

Over the last 15 years, computer-based simulations have emerged as a powerful resource enabling science students to visualize complex phenomena, model dynamic processes, and investigate relationships between elements of physical

Physics Teaching and Learning: Challenging the Paradigm, pages 175–200.
Copyright © 2019 by Information Age Publishing

systems (Clark, Nelson, Sengupta & D'Angelo, 2009; Smetana & Bell, 2012). Properties of simulations such as multiple representations, visual depictions of invisible phenomena, and multiple variables that can be manipulated directly have been found to support students' understanding of physics concepts as well as their participation in forms of exploration similar to scientists' experimental activity (Wieman, Adams & Perkins, 2008). Through efforts like the Physics Education Technology (PhET) project (Perkins et al., 2006), freely available simulations of a wide array of topics and physical phenomena spanning the secondary school and introductory college physics curriculum are readily accessed via internet browsers.

The advent of mobile computing devices such as tablets and smartphones occasions new opportunities for extending the benefits of simulations into new settings and instructional configurations. First, the networking and communication capabilities of personal devices can be integrated with dynamic representation elements of simulations to support potentially novel forms of classroom interaction and peer collaboration (Colella, 2000; Hegedus & Moreno-Armella, 2009; White & Pea, 2011). Second, the touchscreen-based interfaces of contemporary personal digital devices may enable new ways of interacting with simulations, deepening their support for embodied and gestural aspects of reasoning (Lindgren, Tscholl, Wang & Johnson, 2016). Third, sensor capabilities such as accelerometers and magnetometers embedded in many mobile devices can allow the integration of real-world data with virtual aspects of simulations (Blikstein, 2014).

The Physics with Tablets Outside and Inside Classrooms (PHoTOnICs) project is a design-based research project focused on developing and investigating collaborative learning activities by leveraging the capabilities of tablet computers. We have created several tablet and web-based tools and scenarios to support collaborative learning about physics in high school and university classrooms. Each of the PHoTOnICs designs aims to present learners with simulation elements on their own devices they must coordinate with those on the screens of other students in their small group to explore physical phenomena represented by relationships among those elements. These simulations are "distributed by design" (White & Pea, 2011, p. 539) in order to enable new approaches to supporting collaborative learning in physics. Our analyses in this chapter illustrate how these collaborative simulations can provide students with new resources for reasoning about physics concepts.

The focus of this chapter is on the ways in which the distributed and networked nature of our PHoTOnICs learning environments shapes opportunities for reasoning about the relationship between individual and collective physical phenomena. We present two networked simulation designs, targeting the topics of mechanical waves, and electric fields and forces. Through an analysis of episodes of student sense-making about these simulations, we illustrate how the association between individuals and elements of the simulations, and the group's methods of coordinating their interactions with the simulations, become resources for the students to reason about collective physical phenomena.

BACKGROUND

Modern computing tools and communications infrastructures enable rich environments for supporting collaborative learning. Computer-Supported Collaborative Learning (CSCL) environments can vary in the degree they are designed for particular levels of classroom activity. Additionally, the representational infrastructure built into the environment can vary in the degree to which it is designed to support learning a particular content area. For example, the classroom response systems commonly used in university physics classrooms (Dufresne, Gerace, Leonard, Mestre, & Wenk, 1996; Fies & Marshall, 2006) are tailored to whole-class activities but are content agnostic. In contrast, participatory simulation activities are targeted at the whole-class level of activity but tailored to particular content areas (Colella, 2000; Wilensky & Stroup, 1999). These modern technologies contribute to an increasingly social, interactive and dynamic physics classroom. While simulations are likewise highly targeted to learning in particular content areas, however, they are often designed for use by individual learners (e.g., Perkins et al., 2006). The PHoTOnICs project aims to investigate CSCL designs specifically drawing out the social and interactional dimensions of learning in small groups around content-specific simulations.

Teachers often place students in small groups for both practical and pedagogical reasons. Recent national frameworks for science and mathematics education have stressed the importance of providing students with opportunities to organize and express their own thinking, as well as critically examine that of others (National Council of Teachers of Mathematics [NCTM], 2000; National Research Council [NRC], 2012). In this spirit, promoting student participation in conceptually rich classroom discourse has been a central theme in mathematics and science education research and practice over the last two decades (Ball, 1993; Brown & Campione, 1994; Lampert, Blunk & Pea, 1998; Yackel & Cobb, 1996). Research on groupwork among students often distinguishes between cooperative and collaborative learning activities. In cooperative learning activities, students may work in parallel on separate elements of a task. By contrast, collaborative learning is "a coordinated, synchronous activity that is the result of a continued attempt to construct and maintain a shared conception of a problem" (Roschelle & Teasley, 1995, p. 70).

An important component of collaborative learning activities is that students will often articulate their developing understandings to one another, which can stimulate the elaboration of conceptual understandings or create an awareness of the need to revise understandings (Roschelle, 1992). In a study of seventy 4th grade students working on a computer-based task focused on scientific reasoning, for example, Teasley (1995) found pairs of students working on a shared task elaborated on their conceptual talk more frequently than students who talked aloud while working individually. And, in an early and seminal CSCL study, Roschelle and Teasley (1995) conducted a detailed case study of collaboration between two high school students working together on a computer-based simulation of Newtonian mechanics. Through fine-grained analysis of discourse, gesture and interaction, they documented the ways collaborative tasks required the students to make efforts to maintain a shared conception of the problem, verbalize their understandings of the problem and its possible solutions, share knowledge, and monitor one another's understandings (Roschelle & Teasley, 1995). Similarly, Moshkovich (1996) analyzed the collaborative discussions of pairs of students working together in a computer graphing environment. In three case studies, she examined how the students negotiated shared descriptions of the mathematical lines. She found the students' discussions were a rich context for negotiating mathematical meaning, and additionally demonstrated that their negotiations relied heavily on available representational and conversational resources, such as references to objects in the environment, and coordinated talk and gesture. Such multimodal efforts to construct and maintain shared systems of meaning have since been considered a hallmark of collaborative learning.

Collaborative Learning in Networked Classrooms

Several innovative research and design projects recently have begun to map out ways personal devices connected to local computing networks might support and enrich student interaction in mathematics and science classrooms. These networks can allow students to engage in the collective construction of participatory simulations—activities in which each student in a class uses a personal device to control one agent or element in a shared computer-based simulation of a complex system co-constructed by all participants in the classroom group (Colella, 2000; Wilensky & Stroup, 1999). For example, in a detailed examination of a videotaped 90-minute episode from one high school mathematics classroom featuring a classroom network simulation of gridlock phenomena in traffic, Ares, Stroup, and Schademan (2009) analyzed classroom talk and the mediating role of artifacts. They found that classroom networks can support students' agency and participation in collective mathematics activity. In additional investigations of participatory simulation activities in eight socioeconomically diverse urban high school mathematics classrooms, Ares used classroom observations and student interviews to show how these tools also afforded opportunities for students to draw on diverse cultural and linguistic resources for participating in classroom discourse (Ares, 2008). And in studies of student discourse and interaction in high school algebra classes featuring classroom network systems, Hegedus and colleagues have documented the ways these tools facilitate students' attention to and identification with dynamic mathematics representations (Hegedus & Moreno-Armella, 2009; Hegedus & Penuel, 2008).

The classroom network systems and participatory simulations described in these studies were primarily used in learning activities, interactions and discussions among a whole classroom group of students, facilitated by a teacher. However, other research has explored the potential of similar networked personal devices systems to support collaborative learning among smaller groups of students. For example, Zurita and Nussbaum (2004) used a classroom network of handheld computers with a group of 48 first graders in Chile to support collaborative learning activities in mathematics and language. They used pre-and post-tests of student learning, as well as observations of peer collaboration and student interviews, to compare collaborative learning processes and outcomes both with and without the classroom network tools. They found that the networked handheld devices facilitated greater communication, coordination and negotiation among peers. Similarly, White (2006) used a mixed-methods design to investigate student learning outcomes and participation in a heterogeneous middle school mathematics classroom featuring a classroom network system designed to support collaborative mathematics problem solving. Students worked in teams of 4 over a period of three weeks to complete mathematics challenges using networked handheld computers that distributed different representational tools to different student participants. An analysis of pre- and posttest results and video of collaborative problem-solving sessions with eight students in two focus groups found that those students who demonstrated the greatest learning gains were particularly active in using the representational tools as means to participate in the joint problem-solving activities (White, 2006).

Designing Networked Simulations

Simulations are powerful tools for developing conceptual understandings in science, as they can illustrate invisible phenomena, as well as explicitly connect multiple representations of scientific concepts (Blake & Scanlon, 2007; Perkins et al., 2006; Smetana & Bell, 2012). In a review of 61 empirical studies related to the effective use of simulations for science learning, Smetana and Bell found simulations can effectively promote science content knowledge and process skills, and facilitate conceptual change. For instance, in a quasi-experimental study with 90 high-school aged students, Jimoyiannis and Komis (2001) found students who had learned with a Newtonian mechanics simulation performed significantly better on subsequent kinematics reasoning tasks than students in a control group receiving traditional instruction. Many quasi-experimental research designs have shown similar conceptual learning gains related to simulation use within different content areas (McKagan, Handley, Perkins, & Weiman, 2009; Zacharia, 2007; for a review, see Rutten, Van Joolingen, & Van Der Veen, 2012).

Modern personal devices such as tablet computers can provide each student in a classroom with engaging, interactive simulations of scientific phenomena. Our approach to supporting small groups in learning physics is to use both the simulation and networking capabilities of handheld tablets to design group-level tasks in which students work within a shared virtual environment. We aim to design tasks in which the group must truly function as a group to succeed. Such tasks necessitate participation and contributions from each group member, where their contributions are interdependent. The individual touchscreen displays to which each student has access support student interactions with, and reasoning about, representations of physical objects and phenomena, while the networking capability of the devices enables distribution of control over elements of the simulation to support group-level tasks. Each of the PHoTOnICs designs aims to present learners with simulation elements on their own devices they must coordinate with those on the screens of other students in their small group in order to explore physical phenomena represented by relationships among those elements. Such complex group tasks can encourage explicit discussion of the relationships among elements of the simulation, and negotiation of strategies for achieving a shared objective.

Our designs are informed by the principle of generative design (Stroup, Ares & Hurford, 2005): relations between conceptual structure and the social structure of a group can be leveraged to support learning activities. The generative design framework arose from research on learning environments in which students use networked personal devices to participate in mathematics classroom activities. In these scenarios, individual students are mapped by the system to elements of a shared simulation model running on a whole-class display (Hegedus & Penuel, 2008; Hegedus & Moreno-Armella, 2009; Stroup, Ares & Hurford). Similar mappings between conceptual and social structures in classrooms have been produced using classroom networks by White, Wallace, and Lai (2012) in which individual students control and can move points on a Cartesian grid, and a pair of students working together is tasked with producing a line with a particular slope. White et al. performed a detailed examination of one pair of students' collaborative interactions, demonstrating that the networked environment supported the pair in developing coordinated strategies of interacting with their individual points, as well as in establishing mathematical meaning of the relations between their individual points and the collective slope.

The design conjecture behind the collaborative simulations presented in this chapter is that these types of distributed tasks may be well-suited for small groups to learn about particular physics content areas. Just as emergent phenomena are well-suited to investigation in whole-class level participatory simulations, we consider physical systems in which a collective phenomenon is produced by interacting but individual elements may be well-suited for investigations by small groups of students. By distributing control over discrete elements of a system to individual students, and giving the group a shared task to investigate or produce a collective phenomenon, we aim to create a mapping between the social, interactional space of the small group and the conceptual space of the physics under investigation (Stroup, Ares, & Hurford, 2005; White, Wallace, & Lai, 2012). We conjecture that group-level tasks around collective physical phenomena will support student investigation of the relationship between their own actions or interactions with the simulation and the shared (group-level) physics representations. Additionally, we conjecture that the simulations will promote students' reasoning and explicit discussion about the relationships between those individual and collective elements of physical systems.

Reasoning in Networked Learning Environments

Ways of reasoning and participating in mathematical or scientific activity are deeply entwined with the representational and communications infrastructure of the classroom. We should expect small groups to develop discourse and interactional practices specific to any particular design (Roschelle & Teasley, 1995). For example, Stahl (2015) performed detailed discourse analyses of computer chat logs produced by small groups of learners working in an online dynamic geometry environment, Virtual Math Teams. His analyses highlight the complex interplay of discourse particular to mathematics, and forms of practice specific to the dynamic geometry tools (e.g., dragging vertices of a dynamic triangle, as a form of mathematical investigation). Hegedus and Moreno-Armella (2009) similarly illustrated the specificity of discourse practices to the representational tools available. In their classroom network environment for algebra learning, each student in a classroom used a handheld calculator to control a single point on a Cartesian grid, while a network aggregated the points and made them visible on a shared public display. Through a discourse analytic method of examining use of deictic markers as signifiers of identity, Hegedus and Moreno-Armella demonstrated how the classroom network, integrated with the representational infrastructure supporting those particular mathematical activities, created an environment in which learners identified themselves with particular mathematical artifacts (dots on the grid). They argued students' speech acts, as well as physical acts in the classroom, became public enactments of both a personal and mathematical identity. These enactments were observed as a classroom discourse practice in which the students referred to mathematical dots by the name of the student with whom they were associated (e.g., that there were 18 dots, or "19 minus Joe," p. 407).

This referential practice, supported by a mapping between the social and conceptual spaces of the classroom, is similar to that identified in Ochs, Gonzales and Jacoby's (1996) ethnographic and linguistic study of the discourse practices of solid state physicists. The authors analyzed the grammar the physicists used when discussing experimental results during their research group meetings. They found physicists speak in both physics- and physicist- focused terms, but also commonly use *indeterminate references*. Indeterminate references refer simultaneously to a physicist and to a physical object (e.g., "When I come down I'm in the domain state," p. 328). It thus serves as a reference to a "blended identity" (p. 339) blurring the distinction between physicist and physical object.

Similarly to an argument presented by Enyedy, Danish, and DeLiema (2015), we interpret the use of such indeterminate references as evidence of a blending between these social and conceptual spaces. In particular, Enyedy, Danish and DeLiema use the blending between the classroom and conceptual spaces as an analytic lens to characterize reasoning in an augmented reality environment for learning about the physics of forces and friction. In their study, a classroom of second grade students was associated with virtual objects by a display which overlaid the forces on the virtual objects on a video feed of the classroom. Enyedy et al. present a cognitive ethnography of the classroom lesson, documenting the ways this technology configuration supported a blending of the classroom space with a conceptual space of forces and friction. They further demonstrate that this in-between space allowed students to integrate and reason with diverse embodied as well as formal representational resources.

Networked Simulation Designs

In design-based research, theoretical conjectures about the nature of learning are embodied in a learning environment design, and those conjectures are refined in light of an analysis of empirical data (Cobb, et al., 2003; Design Based Research Collective, 2003; Sandoval, 2014). Our conjectures underlying the designs we present here are about the potential of classroom networks of tablets to create the conditions for small groups of students to 1) make sense of dynamic simulations and their interactions with them, 2) establish strategies for coordinating their interactions with these simulations, and 3) articulate conceptual understandings of the underlying physics of the simulations. Particularly, when we distribute control over individual elements of the simulation across students, our conjecture is these designs may be well-suited for small groups to learn about physics related to part-whole or individual-collective relationships.

Below, we present two networked simulation designs for small groups, targeting the topics of 1) electric fields and forces, and 2) mechanical waves. These two content areas were selected for initial networked simulation designs because they both require reasoning about individual-to-collective relations. In both designs presented here, individual students are mapped to individual elements of the physical system to enable group-level investigations of the collective phenomena.

Electric Fields Networked Simulation

Understanding a net field or net force in electricity and magnetism requires understanding how individual charges contribute to the collective construct, and how that contribution depends on the positions and magnitudes of the charges in some spatial configuration. In our *Electric Fields* application, students take responsibility for different charges and different representational resources so they can collectively construct and explore simulations of electric fields and equipotential surfaces. Each student controls the position, sign and magnitude of a single electric charge in a virtual space displayed on the device of each student in the group (see Figures 8.1, 8.2). Students can customize their respective views in a variety of ways, such as showing the net force on their own charge, placing a test charge, drawing a vector, or displaying one or more representations of shared, group-level physical constructs (electric field lines or field vectors, electric poten-

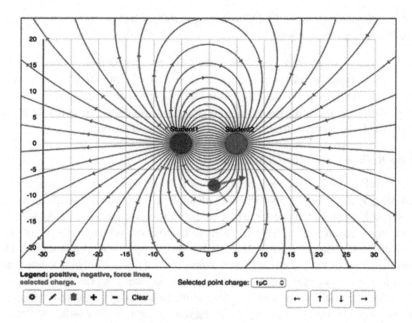

FIGURE 8.1.   Electric Fields application screenshot showing two students' respective charges and electric field lines representation.

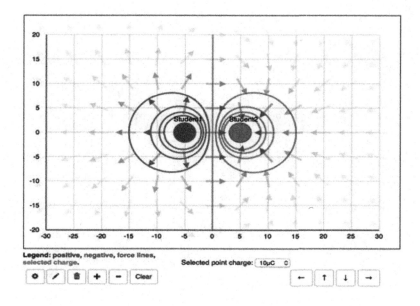

FIGURE 8.2.   Electric Fields application screenshot showing two students' respective charges and vector field representation of the electric field.

tial or equipotential lines). Problem-solving activities supported by this environment include asking students to

- Predict the net force due to other students' charges on their own charge,
- Investigate the relationship between their charge's magnitude, sign or position, and the spatial patterns of the electric field representations, or
- Compare the affordances of different representations for explaining the field associated with a particular configuration of charges.

Wave Builder Networked Simulation

A traveling wave can be thought of as a spatially and temporally coordinated motion of many individual oscillators in a medium. Reasoning about wave phenomena often requires considering both the properties of the collective wave (its frequency, wavelength, or direction of travel) as well as the position, velocity, or phase of any individual point along the wave. In the *Wave Builder* application, each student adjusts the phase of individual oscillators evenly spaced along a horizontal line, and then must work with the group to construct a traveling wave. A primary goal of this simulation is to help students make sense of the relationship between the sinusoidal oscillations, in time, of each point along a wave, and the sinusoidal oscillation across the entire collection of points, in space. To this end we animate a wave as a series of independently oscillating points; each individual point along the wave can

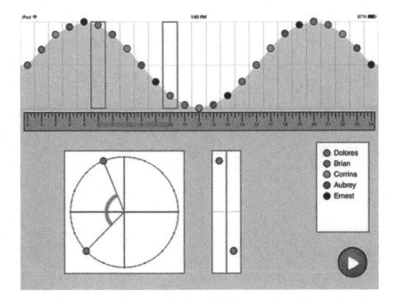

FIGURE 8.3. Screenshot showing one student's view of the Wave Builder application

be selected, and its oscillation represented as a point rotating around a unit circle in time (see Figure 8.3). This is intended to provide a visual representation of the concept of phase or phase angle, which is commonly used in formal, mathematical descriptions of oscillatory motion. Because control over the oscillators' motions is distributed across the students, successfully building a wave requires the students to coordinate with one another to set the motions of their individual oscillators. In sequences of tasks of building waves with different wavelengths and direction of travel, students investigate the relations between the motions of individual oscillators (and phases, as represented on the unit circle), the relative motions of neighboring oscillators, and the collective wave motion.

METHOD

These networked simulation designs were used to support small-group collaborative learning activities in both high school and university classrooms. We investigate how the distributed and collaborative nature of these networked simulation designs can support small group interactions and reasoning about the physics of collective phenomena.

Participants and Data Collection

To investigate the ways these networked simulation designs support collaborative physics reasoning, we purposively selected classroom research sites that already featured regular use of instruction centered on small groups. One of our research sites was an introductory physics course at a large public research university. The course is an exemplar of the reform movement within physics education, in which students are engaged in active learning (Hake, 1998) with a focus on models (Potter et al., 2014). Students in the course spend five hours per week in Discussion/Laboratory sections, led by graduate teaching assistants, working in groups of four to six discussing high-level conceptual physics problems. The second site was the high school classroom of a veteran physics teacher who likewise emphasizes an active learning approach, and who for several years had participated in summer professional development workshops with physics and education faculty involved in physics education reform at the university site. The aim of our research project was to investigate the ways classroom networks of tablets could augment these collaborative learning environments in physics.

In each implementation cycle we brought a classroom set of iPads running the networked simulation software, and set up a local server on a laptop computer connected to a WiFi router. Then, in each class session, we videotaped typically two groups of students from two angles with a pair of digital video cameras as they worked with our applications, for the duration of the activities. As students were already organized into groups of four to six around large, circular tables in the university physics classrooms, and all groups used our activities, we selected to record groups located toward the side of the classroom to minimize the obtrusiveness of our recording equipment. We separately recorded audio using a voice recording application installed on each of the iPads, and synchronized this audio to the video feeds for later transcription of speech. Additionally, researchers present in the classroom documented observations as field notes. We collected student work in the form of worksheets or sketches, as well as photographs of artifacts produced, e.g. on the chalkboards of the university Physics classroom. Further, all students' interactions with the devices broadcast over the network were stored to log files on the server, which were used to reconstruct student device views not captured on camera. Assessments were collected at the university research site in the form of regular course quizzes on the material.

Analytic Approach

As discourse and reasoning practices are enacted not only in verbal utterances, but also through gesture, body language, gaze and interactions with materials, we employed interaction analysis (Derry et al., 2010; Jordan & Henderson, 1995). As our investigation dealt with learning processes, the data we analyze are video records of the small group interactions. Through interaction analysis, repeated viewings of video records enable identification of patterns in discourse or interaction between the participants, and the role student utterances or interactions with the designed materials played in the unfolding group activity at both the macro- and micro-interactional levels.

The video data we analyze below were selected from a broader corpus of over 30 groups, from both a university physics course and a high school physics course, using our networked applications over several rounds of classroom implementations. For this study, we have selected two episodes in which students demonstrated conceptual understandings related to individual-collective relationships. The first episode was selected because we observed a discourse practice similar to that found by Ochs et al. (1996) and Hegedus and Moreno-Armella (2009), as discussed above, that warranted a more detailed examination. The second episode was selected because there was a high degree of verbal interaction and coordination between multiple students, and the group successfully accomplished more than the assigned number wave-building tasks. As every group develops their own ways of talking and reasoning about the simulations, these cases are not intended to be representative of the reasoning done by all of the groups we have studied. Instead, through a fine-grained analysis of these episodes we aim to investigate the ways in which the relational and distributed elements of the collaborative simulations influenced the particular discourse and interactional practices of these groups.

RESULTS

The central idea of this study is that networked learning environments can blend the conceptual domain of physics with the social setting of the classroom in ways that provide students with new resources for reasoning. Both episodes of small group reasoning analyzed below show students' reasoning about physics in ways that reflect the networked and distributed nature of the environment. In the first episode, a group of students using our Electric Fields networked simulation make sense of net forces using a discourse practice reflecting the association between students and charges. In the second, a group using our Wave Builder simulation develops a collective strategy for building waves, and reason about the wave as a collective organization.

Episode One

In this episode, a group of students in an introductory physics course work on an activity supported by the Electric Fields networked simulation. Each student was asked to predict the net force on his or her own charge due to the other students' charges. Each student controlled the position and magnitude of one charge within the virtual environment, and could see the charges of other students in the group on the screen. Whenever a student moved a charge, or adjusted its sign or magnitude, the server updated each simulation with these changes. Making a prediction of net force on any particular charge requires reasoning about the relative positions, signs and magnitudes of the electric charges, and how these independent forces add, as vectors, to produce a net force vector. The students were asked to reproduce their charge configuration on their chalkboard. Then, each student was to make the prediction public by drawing a net force vector on their own charge. The student could check his or her prediction by turning on a setting within the application to display the net force on each charge (see Figure 8.4).

In the following interactional sequence, the students each had drawn their net force vectors on the chalkboard, as arranged in the configuration shown in Figure 8. 4. They were reasoning about the underlying mechanisms explaining the forces on each charge. In this configuration, Andy, Emily and Dora's charges would have a net force toward the bottom-left of the screen, and Cara's toward the top right (as shown in Figure 8.4). As Beth was drawing her net force vector on the chalkboard, Emily noted "Cara is still going in the opposite direction," bringing to the group's attention that Alex, Dora and Emily's positive charges had net forces pointing down-left, while Cara's alone was pointed up-right. The problem became to explain why Cara's charge was behaving apparently differently from the other positive charges.

After Emily problematized the direction of Cara's vector, Cara first tried to explain her net force vector as a repulsion from Emily's nearby positive charge, gesturally highlighting and isolating those two charges and their pair-wise interaction:

Emily: Okay, but, now Cara is still going in the opposite direction
Beth: Yeah, wait what
Cara: well we (pointing to Cara's and Emily's charges on the chalkboard) should be repelling because it's negative

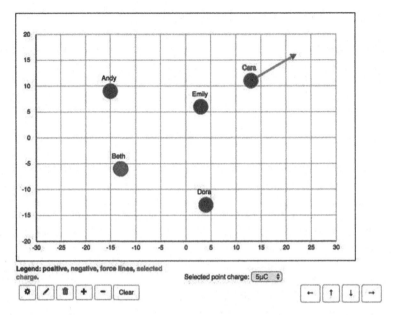

FIGURE 8.4.   E-Fields Student Group Screenshot. Andy, Cara, Dora and Emily control charges of +5 microCoulombs, while Beth has a charge of -5 microCoulombs.

Emily:	(to Cara) Oh, because you're going away
Dora:	you're going away from
Emily:	we're all trying to get away from Beth
Beth:	Wait, no. You guys should be (pointing at Emily's charge on chalkboard)… you (Emily's charge on chalkboard) are trying to get (Beth's charge on chalkboard)
Emily:	I'm still trying to get away from you
Beth:	Well, overall…
Dora:	(to Emily) No, you're trying to go towards Beth; she's (points to Cara) trying to get away from you (points to Emily) cause you're (points to Cara then Emily) in the way of her going to (points to chalkboard) Beth.
Emily:	Oh, right, right, right
Beth:	Because, because we are attracted (pointing to Beth's and Emily's charges), it's just that you have other ones (pointing at Cara's and Dora's charges) that are…

Emily, with "oh, because you're going away," and next a statement that they were "all trying to get away from Beth," suggests Cara's vector is explained as a repulsion from Beth's negative charge. At this point, they are explaining net force vectors with primarily pairwise interactions. Emily's suggestion that they are "all trying to get away from Beth" accomplishes a conceptual shift of the problem to explaining how Cara's positive charge could be attracted to Beth's negative charge, yet still have a net force pointing away from Beth's. Additionally, her suggestion shifts the group's discussion of the problem to explaining where the charges are trying to get. Emily's statement was problematic because it suggested the positive charges should be repelled from Beth's negative charge. Appearing to take note of this inconsistency, Beth disagreed ("wait, no"), pointing to Beth's and Emily's charges on the chalkboard and beginning "you are trying to get..." However, Emily repeated, "I am still trying to get away from you." With "well, overall..." Beth began to shift toward a net, or overall effect, rather than focusing on pairwise interactions. Dora then resolved the problem by explaining: Emily (whose vector was pointing down-left) was trying to go towards Beth; Cara (vector pointing up-right) was trying to get away from Emily, who was "in the way" of Cara getting to Beth. Taking up Emily's framing of the problem, Dora's response both explains the net force vectors in terms of where the charges are trying to get, and incorporates contributions from multiple charges (thus representing a shift away from pair-wise explanations) necessary to explain Cara's net force.

Throughout the episode, the students made use of a particular discourse practice of referring to the charges in the configuration by the name of the person who controls them (e.g., "okay, but now Cara is still going in the opposite direction"). This is similar to the referential practice observed by Hegedus and Moreno-Armella (2009) in which students' names were given to mathematical dots. In both cases, the discourse practice reflects the association enabled by the networked environment between participants and physical or mathematical objects. Often, in our data, we see these sorts of references used to draw attention to locations or directions on the screen relative to these charges (e.g., "move yours closer to Beth"). We consider this to be a simple referential practice in which, for example, Beth becomes a shorthand reference to "Beth's charge." In these cases, the phrases containing the referents do not make relevant any properties of students, or of charges, other than their location, but are consistent with an underlying model in which the students own charges on the screen (e.g., "move yours closer to Beth"). In this episode, however, the student-charges are subjects within a phrase as well as objects (e.g., "she's trying to get away from you"). Here, these references are to constructs having properties of the charges (e.g., a directional vector, or a sign and magnitude of charge), as well as the agentive properties of the students. These references are to blended constructs: part electric charge and part person. We see that Dora, when using such references, gestures either to the students themselves in the classroom or to the charges drawn on the chalkboard, supporting the idea that these are references to such a blended construct.

Much as the physicists studied by Ochs et al. (1996) made use of a blended identity to draw on human cognitive and bodily experience in their collaborative interpretation of graphical representations, we find students in this environment used a similar referential practice to draw on human embodied and social experience in reasoning about physics. Establishing this discourse practice enabled the students to use intuitive, social and embodied resources to reason about the behavior of the student-charges. Their references reflect an underlying model in which the student-charges are attracted to (try to get to) or repel (try to get away from) one another, depending on the signs of the charges. We note their utterances encode a model that works both interpersonally or physically, as either people or charges might be said to repel or attract one another. Dora's statement "you're in the way of her going to Beth," can potentially be due to her ascribing an embodied, social behavior (to be in the way of someone or something) to the charges. Their use of blended constructs appears to establish a blend between these social and physical spaces useful to the students for reasoning about the charge behaviors. For these students, the concept of a net force emerged as something like a balance of social forces: to try to get to someone, but someone else is in the way. As they had been able to explain the net force vectors in this in between space, immediately following this segment, Beth was able to translate it back into physics-focused, formal terms. Thus, the individual associations with particular physical objects in the simulation thus contributed in this episode both to the group's discourse practices and to their reasoning about the physics relating individual charges and net forces. This association supported reasoning in a blended, social-physical space enabling students to use intuitive, embodied resources for reasoning.

Episode Two

As we demonstrated in the previous episode, the association between students and elements of a physical simulation can be reflected in the students' discourse practices. This association also influences the interactional practices by which groups accomplish a shared task. In the following episode, a group of students interacting in the Wave Builder environment establishes local practices of co-ordinating their interactions with the simulations in order to achieve the shared objective of building a wave with a particular wavelength or direction. Across a sequence of wave-building tasks, the students elaborate their strategies and the relationship between their strategy and the resulting wave. We discuss the way this group built their waves, and how their particular practices of sequencing their individual interactions with the tools became a resource for understanding the collective wave.

This was a group of six students: Alex, Bernard, Calvin, Dewey, Elizabeth and Phillip, working in the Wave Builder environment to build a wave with a wavelength of 16. At the start of the activity, Bernard announces that they should each put their points on one of the black dots on the unit circle (see Figure 8.3). As there were 16 of these dots, their strategy would work:

Bernard: So, I'm seeing all these black dots. If we each set at a different black dot we'll get that wave… like there's a black dot here, black dot there (touches each dot in a progression on his iPad)

Agreeing on this strategy of setting points in a progression around the unit circle, the group took ten minutes to build this first wave. Their initial wave-building required a lot of verbal coordination between the students, with Alex, Bernard and Calvin giving instructions to Dewey, Elizabeth and Phillip. In this initial wave-building phase, the students largely gave instructions to one another with specific references to points along the unit circle ("Elizabeth, you'll be at 60," or "Dewey, you're on the wrong dot") or to the selection boxes used to set which point is controlled by the unit circle (e.g., "move your first box to the first yellow dot"). In building this first wave, they frequently leaned over to point to another student's screen to indicate what they needed to do, or rotate their own iPad around to demonstrate. Thus, much of their initial attention was focused on the mechanics of the interactions with the simulation, establishing individual abilities to perform the required actions.

In this group, Bernard frequently took the lead in enacting the group's strategy by announcing whose turn it was or where they should place their point on the unit circle. When breakdowns appeared in enacting the strategy, students occasionally justified a particular next action. In the following segment, the students had continued their counter-clockwise progression of points around the unit circle from zero to 180 degrees. Instead of continuing the progression past 180 degrees, Alex began to place his point again on the black dot before 180 degrees. Taking note of this, Bernard instructed Alex they have to continue to make the "complete circle," and that the wave was now "going to go down":

Bernard: (to Alex) No, no, no, no, we have to keep going, to make the complete circle. It's going to go down now. And, then, that's me next, right?
Calvin: So, now it's going down. On the bottom part of the circle. So, doesn't Phillip stay in that spot?
Bernard: Yeah… then Elizabeth.

This exchange both elaborates the group's strategy (not just to set points in a progression, but to set points in a progression around a complete circle) and explicates the relation between their sequencing on the unit circle (specifically, which quadrant the point is in) and the representation of the wave (whether the points lie above or below the x axis). Additionally, the exchange represents the genesis of a local conception of one wavelength as related to the strategy of building a wave, a full cycle around the unit circle corresponding to completion of one full wave.

Soon after finishing this first wave, the graduate Teaching Assistant (TA) approached the group and asked them what they had done. The group explained they had gone in increments of 30 degrees, and the researcher asked how they had chosen that value.

TA: How is it that you chose 30 degrees, for that value?

Alex: It's just we're going 30, 45, 60...

Bernard: It wouldn't make sense to have it 30, then like 66... (gesturing large jump around the unit circle)

Alex: It gives you a (gestures small delta y) good curve (gestures a crest)...

Bernard: It's incremental.

Alex: If you did like too big of an increment (large motion around a unit circle) it wouldn't (gestures a bigger delta y) look exactly like a wave (gestures curve), too small, I don't know...

TA: Yeah, think about what it would be if you made it too small.

Bernard: It would... cycle a lot (gestures rapid up and down)). No, it would be longer (gestures long crest)... it would be longer. Because it would take longer for... it would take more dots (gesturing along a wavelength) to create the wave. So, it would change the... wavelength?

While their initial strategy of setting points in increments around the unit circle had worked, it had not been explicitly related to the wavelength of the resulting wave. When asked how they chose that increment, they only said the evenness of the spacings should produce a smooth wave. When prompted to explain what would happen to the resulting wave if they had chosen a smaller increment, however, Bernard began to articulate the relationship between their increments in their strategy and the resulting wavelength of the wave, that "it would take longer... it would take more dots" to create the wave. Later, when asked by Elizabeth to re-explain this relationship, Bernard drew sketches of a unit circle and a graphical representation of a wave to explain how the wavelength would change if they had used a smaller increment:

Bernard: So, it becomes like (stands and turns to chalkboard)... So, if you have... if you have a graph (draws graph of a wave, labels points on x axis from 0 to 2Pi, draws two cycles of a wave)... she was saying if you're doing this (makes three tick marks on the unit circle between 0 and 90) which is what we were doing, right? Then, we're just going to go like one, two, three, four, five (drawing points along the first half of the wave graph). But, if we made it so that there was more (adding tick marks between each of the previous drawn on the unit circle)... if we made it so that there was more, then this... (gesturing to the first crest of the graph and motioning to the right) this would increase... this (gestures an interval between dots with his finger and thumb, expands it) wavelength would increase. Because it takes a longer time to [[inaudible]]. Which is what this (points to iPad on the table) is doing, except we were going in increments of 30 degrees, right?

Bernard's explanation is an explicit relating of the group's sequencing on the unit circle to the wavelength of the resulting wave. They then began to see a relationship between the spacings around the unit circle in their strategy and the resulting wavelength, and more specifically, changes in their strategy corresponding to changes in wavelength. After representing the relationship between their strategy and the wavelength of their wave in this way, the group appeared to now understand how to modify their strategy to build waves of different wavelengths. They next built two waves, one with a wavelength of 32, and another of 8. They built their wave with a wavelength of 32 by halving the increment on the unit circle in their original strategy. While working through building this wave, Bernard again articulated a relation between their strategy and the wavelength of the resulting wave: "It's just taking us longer to get around the unit circle, which is increasing our wavelength," suggesting a wavelength corresponds to a full cycle around the unit circle. After finishing this wave, they decided to build one with a wavelength of 8, which they were able to do with very little verbal coordination. When asked, in a whole class discussion, how they built this last wave, Bernard articulated this as a modification of their initial strategy: "The last one was 16 and we wanted to make it 8, so we had to double the distance. So, instead of every black dot, every other." Their initial strategy used to build a wavelength of 16 became a resource for their building of other wavelengths; the increment used in their original strategy was adjusted to build longer or shorter wavelengths.

The final task in the activity sequence was to reverse the direction of the wave, which gave the group a new opportunity to consider the relationship between their sequencing on the unit circle and the characteristics of the resulting wave, this time to its direction. They struggled with how to modify their strategy to reverse the wave's direction: whether they should shift every point by some angle around the unit circle, or flip their points around either the x or y axes. Groups reasoned about this task in different ways, which often appeared to depend on the ways in which they initially conceived of their wave building (as setting points around the unit circle, or as using the unit circle to set the positions and directions of oscillators along the wave). This group ultimately decided to reverse the direction of their original progression around the unit circle from counter-clockwise to clockwise. Though this new strategy worked, they struggled to explain why. They attempted to think of the reversal of the progression in terms of going in either a positive or negative direction around the unit circle, but weren't able to make sense of how this should relate to the wave direction. The Teaching Assistant later drew their attention to the relative motions of two neighboring oscillators, which then allowed them to explain how the direction of their progression around the unit circle produced a particular direction of motion. Just prior to that final discussion, Bernard explained his understanding of a wave as a collective organization of individual oscillators to the group:

Bernard: To me, it sounds like unit circle indicates the direction of the wave, like... together. Separately, it only indicates position. It's like an x-y point, right? Like all together it is a line, but individually it's just a point, right? So, based on the collection of points on the unit circle... will be the direction of the wave.

Our design conjecture about mapping the social and conceptual physics spaces using networked technologies appeared here as Bernard's articulation of the relationship between individual oscillators' motions and the direction of the collective wave. By drawing an analogy to a line as a collection of individual points, he demonstrates an understanding of wave direction as a collective phenomenon arising from independent oscillators' motions.

DISCUSSION

Together, our analyses illustrate how an integrated communications and representational infrastructure integrate the social and interactional space of the classroom with the conceptual space of physics. In our first analysis, we illustrated how a mapping between participants and elements of the simulation supported reasoning in a blended space of student-charges, reflected in the group's discourse practices as indeterminate references to a blended identity (Ochs, Gonzales, & Jacoby, 1996). A similar referential practice in mathematics classrooms was interpreted by Hegedus and Moreno-Armella (2009) as a new form of discourse enabled by classroom networks, in which students' enactments of their personal identities in the classroom became imbued with mathematical meaning as the students adopted and enacted these blended identities. Here, we highlight the effect this association had on the students' reasoning about physics: the indeterminate references reflected reasoning in a blended space of student-charges inviting the use of social-embodied resources for understanding the net forces on charges. This result parallels work done by Enyedy, Danish and DeLiema (2015) in augmented reality environments. Though the forms of technological support are quite different, both created an association between learners and elements of physical simulations. In both cases, a blend between the social and physical spaces resulted, broadening the resources available for students to reason about the physical phenomena.

We also illustrated a blending between the interactional and conceptual spaces in our second analysis. In this case, the networked simulations allowed for a series of collaborative tasks requiring the group to develop and elaborate schemes of interaction with the tools. Specifically, the group developed a collective strategy for coordinating their interactions with the simulation that became a resource for reasoning about collective properties of waves. The group was able to explicitly relate aspects of their strategy (the increment and direction of progression) to aspects of the wave (wavelength, direction). This represents a different understanding of wave concepts than is typically achieved in physics courses, though it targets the same conceptual relations. Physics courses are often focused on graphical or symbolic representations of waves, and use simulations as a way to dynamically represent wave phenomena and make connections to these formal representations. In our *Wave Builder* environment, the students came to understand the wave as collectively built through a coordination of individual oscillators, and to see the relationship between the aspects of that coordination (spacing around the unit circle, and the direction of that progression) and the wavelength and direction of the collective wave.

Both of the designs presented leveraged a congruence between the interactional space of the small group and the conceptual space of physics, in which individual-group relations mirrored the individual-collective relations of electric charges and fields, or oscillators and waves. These learning activities illustrate how Stroup, Ares and Hurford's (2005) principles of generative design can be extended from mathematics classrooms to physics simulations. The networking capability of the devices enabled associating individuals with elements of the physical system, and distributing control over those elements. Mapping the social structure of a collaborative group to these corresponding physical relationships in this way allowed each student to have a particular role in the creation of a net force, or in the building of a wave. The individual displays to which each student had access supported reasoning about representations of these physical systems and created a shared context in which students had both a common problem-solving goal and a shared set of tools with which to solve it. Moreover, distributing control over elements of the simulation across multiple students' devices promoted a positive interdependence of student contributions, requiring the group to coordinate their interactions to successfully complete the shared tasks (Webb & Palincsar, 1996; White & Pea, 2011). This provided students with opportunities to consider the nature of their own contribution to the collective constructs. Further, as collaborative tasks require students to make efforts to maintain a shared conception of a problem, and encourage verbalizing their understandings of the problem and its possible solutions (Roschelle & Teasley, 1995), the relation between individual elements and the collective physical constructs could become explicitly discussed.

In our analyses we illustrated two ways in which the relation between group and physical structure may influence student reasoning. In our first analysis, the association between students and electric charges supported a conceptual blending between the student space and the conceptual space of electric charges influencing the ways in which the students both referred to and reasoned about the behavior of a collection of charges (Enyedy, Danish & DeLiema, 2015; Hegedus & Morena-Armella, 2009). In the second analysis, the shared task required the group to develop a strategy for building waves in which they sequentially coordinated their interactions with the simulation. The simulation acted as a context to talk about or act in (Roschelle & Teasley, 1995), but also as a context in which to establish this collective strategy. The students appeared to understand waves as a collective phenomenon, where aspects of their wave-building strategy mapped to properties of a resulting wave. By intertwining these often distinct social and conceptual spaces, relations and coordinations between group members became physically meaningful as resources for understanding relations between individual and collective physical constructs.

CONCLUSION AND IMPLICATIONS

As physics educators incorporate modern technologies into physics classrooms in support of active and collaborative learning environments, physics classrooms can become more social, interactive and dynamic. While simulations have been shown to be powerful instructional tools, they are not often designed to take into account the social structure of the classrooms in which they are being used. In line with principles of generative design (Stroup, Ares & Hurford, 2005), we suggest simulations can be intentionally designed to be used by small groups, by giving groups a shared virtual environment in which each group member can interact, discuss and collaboratively reason about shared phenomena. Networked simulation designs can integrate the social setting of learning with conceptual content, capitalizing on peer interaction to provide new opportunities and resources for students to reason about physics.

In agreement with the principle of generative design (Stroup, Ares, & Hurford, 2005), our results also suggest collaborative learning in small groups may be a particularly suitable instructional configuration for learning about interdependence or relations between elements of physical systems. By distributing control over elements of a simulation, students are provided with opportunities to investigate and explicitly discuss their individual effects on a collective phenomenon, or the ways in which collective actions can produce a particular physical phenomenon. Simulations targeting small groups can make use of this congruence between the part-whole relationships of both the social and conceptual spaces to expand available ways to interact and reason about physical concepts.

Future research will pursue additional ways in which simulation designs may promote the integration between social and conceptual dimensions of physics reasoning, and identify content areas in which this may be particularly beneficial for student understanding of and reasoning about physics concepts. That is, under what conditions is a blending between the social and conceptual spaces productive for conceptual development? Future research may also seek to relate participation in these particular ways of talking and reasoning in networked learning environments to performance on assessments of conceptual learning, such as on concept inventories. As the episodes we focused on in this study featured little involvement of a teaching assistant or instructor, additional research will be necessary to understand the role of the instructor in guiding small group learning as supported by networked simulations, and what forms of instructional support are needed to maximize learning gains from such designs for small group collaborative learning. More important may be continued research into how use of these sorts of collaborative activities based on networked simulations may contribute to an active and social culture of learning in physics classrooms.

ACKNOWLEDGEMENTS

This material is based upon work supported by the National Science Foundation under Grant No. 1252508. We dedicate our paper to the memory of Wendell Potter, who was a partner in the design of the learning activities presented here and whose pioneering efforts to foster student interaction and collaboration in physics classrooms were an inspiration for this project.

REFERENCES

Ares, N. (2008). Cultural practices in networked classroom learning environments. *International Journal of Computer-Supported Collaborative Learning, 3*(3), 301–326.

Ares, N., Stroup, W. M., & Schademan, A. R. (2009). The power of mediating artifacts in group-level development of mathematical discourses. *Cognition and Instruction, 27*(1), 1–24.

Ball, D. L. (1993). With an eye on the mathematical horizon: Dilemmas of teaching elementary school mathematics. *The Elementary School Journal, 93*(4), 373–397.

Blake, C., & Scanlon, E. (2007). Reconsidering simulations in science education at a distance: Features of effective use. *Journal of Computer Assisted Learning, 23*(6), 491–502.

Blikstein, P. (2014). Bifocal modeling: Promoting authentic scientific inquiry through exploring and comparing real and ideal systems linked in real-time. In A. Nijholt (Ed.), *Playful user interfaces* (pp. 317–352). Singapore: Springer.

Brown, A. L., & Campione, J. C. (1994). Guided discovery in a community of learners. In K. McGilly (Ed.), *Classroom lessons: Integrating cognitive theory and classroom practice* (pp. 229–270). Cambridge, MA: The MIT Press.

Clark, D. B., Nelson, B., Sengupta, P., & D'Angelo, C. M. (2009). *Rethinking science learning through digital games and simulations: Genres, examples, and evidence.* Paper commissioned for the National Research Council Workshop on Games and Simulations, Washington, DC.

Cobb, P., Confrey, J., DiSessa, A., Lehrer, R., & Schauble, L. (2003). Design experiments in educational research. *Educational Researcher, 32*(1), 9–13.

Colella, V. (2000). Participatory simulations: Building collaborative understanding through immersive dynamic modeling. *The Journal of the Learning Sciences, 9*(4), 471–500.

Design-Based Research Collective. (2003). Design-based research: An emerging paradigm for educational inquiry. *Educational Researcher, 32*(1), 5–8.

Derry, S. J., Pea, R. D., Barron, B., Engle, R. A., Erickson, F., Goldman, R., ... Sherin, B. L. (2010). Conducting video research in the learning sciences: Guidance on selection, analysis, technology, and ethics. *Journal of the Learning Sciences, 19*(1), 3–53.

Dufresne, R. J., Gerace, W. J., Leonard, W. J., Mestre, J. P., & Wenk, L. (1996). Classtalk: A classroom communication system for active learning. *Journal of Computing in Higher Education, 7*(2), 3–47.

Enyedy, N., Danish, J. A., & DeLiema, D. (2015). Constructing liminal blends in a collaborative augmented-reality learning environment. *International Journal of Computer-Supported Collaborative Learning, 10*(1), 7–34.

Fies, C., & Marshall, J. (2006). Classroom response systems: A review of the literature. *Journal of Science Education and Technology, 15*(1), 101–109.

Hake, R. R. (1998). Interactive-engagement versus traditional methods: A six-thousand-student survey of mechanics test data for introductory physics courses. *American Journal of Physics, 66*(1), 64–74.

Hegedus, S. J., & Penuel, W. R. (2008). Studying new forms of participation and identity in mathematics classrooms with integrated communication and representational infrastructures. *Educational Studies in Mathematics, 68*(2), 171–183.

Hegedus, S. J., & Moreno-Armella, L. (2009). Intersecting representation and communication infrastructures. *ZDM, 41*(4), 399–412.

Jimoyiannis A., & Komis V. (2001). Computer simulations in physics teaching and learning: a case study on students' understanding of trajectory motion. *Computers and Education, 36*, 183–204.

Jordan, B., & Henderson, A. (1995). Interaction analysis: Foundations and practice. *The Journal of the Learning Sciences, 4*(1), 39–103.

Lampert, M., Blunk, M. L., & Pea, R. (Eds.). (1998). *Talking mathematics in school: Studies of teaching and learning*. New York, NY: Cambridge University Press.

Lindgren, R., Tscholl, M., Wang, S., & Johnson, E. (2016). Enhancing learning and engagement through embodied interaction within a mixed reality simulation. *Computers & Education, 95*, 174–187.

McKagan, S. B., Handley, W., Perkins, K. K., & Wieman, C. E. (2009). A research-based curriculum for teaching the photoelectric effect. *American Journal of Physics, 77*(1), 87–94.

Moschkovich, J. N. (1996). Moving up and getting steeper: Negotiating shared description of linear graphs. *The Journal of the Learning Sciences, 5*(3), 239–277.

National Council of Teachers of Mathematics (Ed.). (2000). *Principles and standards for school mathematics* (Vol. 1). National Council of Teachers of Mathematics. Reston, VA: Author.

National Research Council. (2012). *A framework for K–12 science education: Practices, crosscutting concepts, and core ideas*. Washington, DC: National Academies Press.

Ochs, E., Gonzales, P., & Jacoby, S. (1996). "When I come down I'm in the domain state": Grammar and graphic representation in the interpretive activity of physicists. *Studies in Interactional Sociolinguistics, 13*, 328–369.

Potter, W., Webb, D., Paul, C., West, E., Bowen, M., Weiss, B., ... De Leone, C. (2014). Sixteen years of collaborative learning through active sense-making in physics (CLASP) at UC Davis. *American Journal of Physics, 82*(2), 153–163.

Perkins, K., Adams, W., Dubson, M., Finkelstein, N., Reid, S., Wieman, C., & LeMaster, R. (2006). PhET: Interactive simulations for teaching and learning physics. *The Physics Teacher, 44*(1), 18–23.

Podolefsky, N. S., Perkins, K. K., & Adams, W. K. (2010). Factors promoting engaged exploration with computer simulations. *Physical Review Special Topics-Physics Education Research, 6*(2), 1–11.

Roschelle, J. (1992). Learning by collaborating: Convergent conceptual change. *The Journal of the Learning Sciences, 2*(3), 235–276.

Roschelle, J., & Teasley, S. D. (1995). Construction of shared knowledge in collaborative problem solving. In C. O'Malley (Ed.), *Computer-supported collaborative learning* (pp. 69–97). New York, NY: Springer.

Rutten, N., Van Joolingen, W. R., & Van Der Veen, J. T. (2012). The learning effects of computer simulations in science education. *Computers and Education, 58*(1), 136–153.

Sandoval, W. (2014). Conjecture mapping: An approach to systematic educational design research. *Journal of the Learning Sciences, 23*(1), 18–36.

Smetana, L. K., & Bell, R. L. (2012). Computer simulations to support science instruction and learning: A critical review of the literature. *International Journal of Science Education, 34*(9), 1337–1370.

Stahl, G. (2015). *Constructing dynamic triangles together: The development of mathematical group cognition.* Cambridge, UK: Cambridge University Press.

Stroup, W. M., Ares, N. M., & Hurford, A. C. (2005). A dialectic analysis of generativity: Issues of network-supported design in mathematics and science. *Mathematical Thinking and Learning, 7*(3), 181–206.

Teasley, S. D. (1995). The role of talk in children's peer collaborations. *Developmental Psychology, 31*(2), 207–220.

Webb, N. M., & Palincsar, A. S. (1996). Group processes in the classroom. In D. C. Berliner & R. C. Calfee (Eds.), *Handbook of educational psychology* (pp. 841–873). New York, NY: Macmillan Library Reference USA; London, England: Prentice Hall International.

White, T. (2006). Code talk: Student discourse and participation with networked handhelds. *International Journal of Computer-Supported Collaborative Learning,1*(3), 359–382.

White, T., & Pea, R. (2011). Distributed by design: On the promises and pitfalls of collaborative learning with multiple representations. *Journal of the Learning Sciences, 20*(3), 489–547.

White, T., Wallace, M., & Lai, K. (2012). Graphing in groups: Learning about lines in a collaborative classroom network environment. *Mathematical Thinking and Learning, 14*(2), 149–172.

Wieman, C. E., Adams, W. K., & Perkins, K. K. (2008). PhET: Simulations that enhance learning. *Science, 322*(5902), 682–683.

Wilensky, U., & Stroup, W. (1999). Learning through participatory simulations: Network-based design for systems learning in classrooms. *Proceedings of the 1999 Confer-*

ence on Computer Support for Collaborative Learning, (1), 80. Retrieved from http://portal.acm.org/citation.cfm?id=1150240.1150320

Yackel, E., & Cobb, P. (1996). Sociomathematical norms, argumentation, and autonomy in mathematics. *Journal for Research in Mathematics Education, 27*(4), 458–477.

Zacharia, Z. C. (2007). Comparing and combining real and virtual experimentation: An effort to enhance students' conceptual understanding of electric circuits. *Journal of Computer Assisted Learning, 23*(2), 120–132.

Zurita, G., & Nussbaum, M. (2004). Computer supported collaborative learning using wirelessly interconnected handheld computers. *Computers & Education, 42*, 289–314.

CHAPTER 9

DESIGN, IMPLEMENTATION AND EVALUATION OF A RESEARCH-INFORMED TEACHING SEQUENCE ABOUT ENERGY

Dora Orfanidou and John Leach

We present the design, implementation and evaluation of a research-informed teaching sequence that aims to promote conceptual understanding about energy among Cypriot students, ages 15–16. The design addressed the requirements of the official curriculum and was underpinned by Feynman's (Feynman, Leighton, & Sands, 1963) abstract perspective. Two design tools were used to inform design decisions: learning demand and communicative approach. A quasi-experimental design was used to compare the conceptual understanding of students in a group (n=18) who had followed the designed teaching sequence, with a comparison group (n=18) who had followed the school's usual approach. Students who were taught through the designed teaching sequence were significantly better at producing energy descriptions of physical systems than were students who had followed usual traditional teaching, demonstrating that a conceptual treatment of energy can be accessible to students at this age and stage of their physics education.

Physics Teaching and Learning: Challenging the Paradigm, pages 201–246.
Copyright © 2019 by Information Age Publishing
201

Keywords: design research, teaching energy, research-informed design, conceptual approach to energy, visual representations, evaluation of teaching

This chapter addresses a core question in physics education at the high-school level: *is it possible to introduce physics learners to an account of the energy concept that they can use successfully, and that is consistent with contemporary conceptual perspectives on energy?* Several curriculum designs are described in the literature which present a quasi-material account of energy, on the grounds that this is a productive first step for students towards a more sophisticated and conceptual approach to understanding energy. By conceptual approach we mean an approach giving primacy to the ontology of energy as a numerical value that remains the same through change processes. Such approaches begin with the assumption that a conceptual approach is inaccessible to students at the introductory level. We set out to investigate this assumption by carrying out a design study to evaluate whether it is feasible to introduce a comprehensible, usable perspective on energy to students consistent with contemporary conceptual perspectives on energy.

The chapter begins by setting out the design, implementation and evaluation of a research-informed teaching sequence aiming to promote conceptual understanding about energy among physics students ages 15–16, following the Cyprus physics curriculum. The rationale for the design is presented and compared to other approaches reported in the literature. As well as meeting the requirements of the Cyprus National Curriculum, the design specification also emphasized internal coherence and consistency with contemporary perspectives on energy. A perspective on energy was therefore taken consistent with Feynman's (Feynman et al., 1963) abstract perspective where conservation is the key property. Design decisions about the treatment of content and teacher-to-whole-class talk were informed by a perspective on the design of science teaching set out by Leach, Scott and associates (Leach & Scott, 2003). Design decisions are informed using both theoretical and empirical insights at a fine grain size. Two design tools were used to inform design decisions, namely learning demand (to inform content-related design decisions; Leach & Scott, 2002) and communicative approach (to inform decisions about teacher talk in the classroom; Mortimer & Scott, 2003).

The chapter goes on to describe the implementation of the teaching sequence, to evaluate its success at achieving its aims, and to discuss the implications from this work for other designers of teaching about energy. A quasi-experimental design was used to compare the conceptual understanding of students in a group (n=18) who had followed the designed teaching sequence, with a comparison group (n=18) who had followed the school's usual and traditional approach to teaching. The students in both groups were comparable in terms of prior attainment in physics, and the teachers were comparable in terms of qualifications, motivation and experience. Evidence is presented that the designed teaching sequence was successful in enabling students to use the aspects of store, transfer, conservation and degradation to produce energy descriptions of physical systems. Experimental students were significantly better at using the elements to produce energy descriptions than were students who had followed the school's usual teaching approach. We claim a conceptual treatment of energy can be accessible to students at this age and stage of their physics education. The chapter concludes with a discussion of the kinds of claims that can be supported from design research such as this one.

PERSPECTIVE ON ENERGY TAKEN FOR DEVELOPMENT OF THE RESEARCH-INFORMED TEACHING SEQUENCE

The physicist Richard Feynman formulated a theoretical framework for energy as follows:

> There is a fact, or if you wish, a *law*, governing all natural phenomena that are known to date. There is no known exception to this law-it is exact so far as we know. The law is called the *conservation of energy*. It states that there is a certain quantity, which we call energy, that does not change in the manifold changes which nature undergoes. That is a most abstract idea, because it is a mathematical principle; it says that there is a numerical quantity which does not change when something happens. It is not a description of a mechanism, or anything concrete; it is just a strange fact that we can calculate some number and when we finish watching nature go through her tricks and calculate the number again, it is the same. (Feynman, Leighton & Sands, 1963, p. 4–1)

The theoretical framework developed on which the research-informed teaching sequence for energy was grounded was underpinned by the Feynman's perspective. Specifically, this theoretical framework emphasizes the abstract nature of energy; it is not anything concrete. The framework further emphasizes the key property of the concept: conservation of energy. Our intention was to propose a teaching sequence for energy based, as closely as possible, to the abstract and conservable nature of energy, in line with Feynman's position and current scientific beliefs. We are aware of the difficulties in developing an accessible teaching sequence for students in secondary education focusing on the abstract and mathematical nature of energy, discussing such difficulties at various parts of following sections. Feynman used a materialistic analogy to explain this most abstract idea by comparing energy to a child's blocks. The use of a materialistic analogy, however, did not imply energy could be considered as a substance-like entity. Our decision to choose to develop our theoretical framework on the conceptual nature of energy and not on a materialistic metaphor is further supported by the concluding sentences of Feynman's lecture: "What is the analogy (the child's blocks) of this, to the conservation of energy? The most remarkable aspect that must be abstracted from this picture is that there are no blocks!" (Feynman, Leighton & Sands, 1963, p. 4–2).

LITERATURE ABOUT THE ENERGY CONCEPT FOR THE PURPOSE OF TEACHING

In a series of articles, Warren (1982, p. 295; 1983; 1986; 1991) argues for a conceptual view (1982, p. 295) of energy within the context of school physics; an approach consistent with Feynman's notion of energy as a mathematical abstraction. According to Warren, (1991, p. 8) energy is a mathematical abstraction "which is the capacity of a body or system for doing work (in the scientific meaning of the word) by ideal processes." In turn, Schmitt (1982) reported on a theoretical framework developed by Falk and Herrmann (1977, 1978, 1979, 1981) in which energy is seen through a "materialist" (Warren, 1982, p. 295) perspective, in contrast to Feynman's position. Specifically, energy is considered as a substance-like quantity which can be stored in a system and flow in a particular form from one system to another. The flow of energy is always accompanied by the flow of at least one other substance-like quantity, which carries energy.

There are several examples where a materialist approach has been used in the teaching of energy in the research literature, on the grounds that this approach makes the energy concept accessible to students. For example, a substance-like conception of energy, and the idea of energy flow were used by Duit (Duit, 1985; Duit & Haeussler, 1994) for the development of an instructional approach. In this approach, energy is characterized by five basic aspects: conceptualization (in effect the fundamental way in which energy is conceptualized), transfer, transformation, conservation and degradation. Furthermore, energy is conceptualized as "a general kind of fuel" (Duit, 1985, p. 93) needed in order for physical and technological processes to take place. Another instructional approach presenting energy as substance-like was proposed by Millar (2005, p. 6): .".. it is hard not to think of energy as 'something' that flows, or is somehow transferred, from place to place than just thinking of it as a number that does not refer to anything "real".'

In a research report, Scherr, Close, McKagan, and Vokos (2012) focused on a substance metaphor of energy, claiming the advantages of its use in the teaching of energy outweigh any concerns. The researchers claimed their substance metaphor approach facilitates understanding of the key properties of conservation, transfer and flow of energy. The use of a substance metaphor to teach the concept of energy to introductory college or university physics courses was also proposed by Brewe (2011) who claims the utilization of a substance metaphor as a central feature in introductory physics courses provides students with a rich set of conceptual recourses to reason about conservation, storage and transfer of energy. As he states, the use of the substance metaphor alone cannot promote energy as a viable way of modelling physical systems. This modelling can be accomplished with the incorporation of multiple tools for representing and reasoning about energy conservation, storage and transfer.

The substance metaphor for teaching energy was used by Gupta, Hammer and Redish (2010) in which the appropriateness of ontological blending of physical concepts was discussed. As Gupta et al. claimed, the conceptualization of nonsubstance concepts using substance-based ideas is part of expert physicists' language and reasoning in scientific contexts. They suggested physics instruction should enable students to build expert concepts on their everyday resources reinforcing the expert-like thinking skills as these appear in classroom discourse. They emphasized, however, that substance-based reasoning entails the risk that naive reasoning dominates in learners' explanations of physics contexts, demonstrating incorrect reasoning. In contrast to Brewe's and Gupta et al.'s views, Slotta and Chi (2006) state: "Our research suggests that instruction should stress the basic ontological characteristics of the concepts, targeting students' existing conceptions indirectly by carefully avoiding any language, analogies, or phenomenon that might otherwise reinforce the substance-based view" (pp. 286–287). We designed the intervention used in this study on the same basis as Slotta et al.'s position that physical concepts should be introduced to students through teaching approaches which would be underpinned by a theoretical framework which should be in accordance with, or as close as possible to, the ontology used by scientists in explanations. We were interested to find out to what extent it was possible to enable secondary physics students to use a conceptual account of energy without recourse to intuitive materialistic conceptions. We recognize that physics experts have the ability to traverse ontologies of physical concepts in their professional contexts and still understand the actual ontology of each concept. The designed teaching intervention assumed an understanding of the ontology of the energy concept by students might be hindered by the use of substance-like metaphors.

In a recent study, Yao, Guo, and Neumann (2017) report the results of a refined learning progression of energy developed in a previous study. The study aimed to investigate students' progress in developing understanding from an everyday account of energy to a more scientifically consistent one. The learning progression of energy was built across two dimensions: a) the use of the aspects of forms, transfer, transformation, dissipation and conservation and b) the use of four conceptual levels, namely fact (students describe, and interpret daily life phenomena using everyday experience and piece knowledge, which are unconnected to each other); mapping (students can articulate the concept by mapping its abstract feature to observable physical quantities); relation (students can articulate relationships between several concepts or specific mechanisms); and systematic (students can coordinate more than one concept in multivariate systems in a variety of contexts). The refined learning progression of energy was implemented among 4550 students from grades 8 to 12 from two districts in a major city of mainland Peoples Republic of China. Results of data analysis did not support the researchers' hypothesis that students' progress along this sequence of distinct, hierarchically ordered conceptions in their understanding of energy. Furthermore, results confirmed the researchers' second hypothesis that students' progress in their understanding of energy conceptual development levels and in particular, this increases at high school levels.

The literature, however, includes other examples where a conceptual approach is used. The rationale for such schemes is to present to learners a view of energy more consistent with that used in physics, while being as accessible to students as possible. For example, Ogborn claimed '.'..energy is *not* the 'go' of things, despite the common belief that it is. That is, possession of energy, is *not* what drives, gives potential for, explains, or accounts for change" (1986, p. 30). The occurrence of changes should be attributed to another physical concept, which is very close to that of a fuel, named entropy or free energy (Ogborn). Solomon (1982) remarked concepts like entropy and free energy and the Second Law of Thermodynamics are quite difficult and abstract to be understood by students as well as to be introduced in a comprehensive manner by science teachers. She suggested instruction could be possible using simpler and more familiar terminology. In a later piece of work, Ogborn (1990) suggested the use of the idea of differences to replace the concept of negative entropy and free energy for the interpretation of the observed changes. Within this theoretical framework, it is considered that in a closed system, there are limits as to what can happen and what cannot. One such limit is that total energy of the system cannot change. These ideas were used by Boohan and Ogborn (1996) for the development of an instructional approach for teaching energy and chemical change with lower secondary school students, using visual representations of the change phenomena. This approach was evaluated by Stylianidou (1997) with three 11-year-old students who were reported to have worked successfully with the activities, making use of the targeted terms and representations. The abstract pictorial representations seemed to contribute to the achievement of higher levels of generalization in the students' explanations of physical, chemical and biological changes than might otherwise have been expected. The teaching, however, required students to get to grips with a different set of fairly complex representations for each change process, thereby reducing the likelihood they would develop a view of energy as a generalizable and unifying concept.

A research-based learning progression of energy from primary to upper secondary school was reported by Colonnese, Heron, Michelini, Santi, and Stefanel (2012) which they named "Energy Vertical Path" (p. 45). According to the researchers, the Path aimed to reconstruct students' pre-instructional views about energy towards a scientifically oriented conception of energy which could be extended and refined in future teaching. At the primary school level, the concept of energy was introduced as a state property of objects, occurring in four types, namely, kinetic energy, potential energy, internal energy and energy associated with light. These types can be transformed from one to another through interactions. At the lower secondary school level, transformations were treated quantitatively using data from simple laboratory activities. Finally, at upper secondary school level the conversion of the different types into internal energy and the aspect of conservation were introduced. The primary and the lower secondary school levels of the Path were pilot tested whereas the upper secondary school level was explored in an unsystematic way. For the evaluation of the first two levels, pre- and post-tests and interviews were used. The researchers claimed their results provide evidence of an important improvement of energy ideas among the majority of the participant students, though no results were reported at the upper secondary school level.

The use of visual representations to teach the energy concept was also proposed within the Supporting Physics Teaching 11–14 (SPT11-14) project (Institute of Physics, 2006). Energy was treated as a unifying concept, abstract, calculable and conserved in nature. Energy can be found in energy stores defined "… as places where one can pin down quantities of energy by calculation" (Lawrence, 2007, p. 403). Depending on the calculating mechanism, energy can be found in various different stores, represented as tanks which can be filled or emptied. We recognize this representation, in itself, is materialist. The point made by Millar (2005) indicating finding a language to talk about energy that does not portray it as something is difficult. Later in the chapter we indicate the measures we took to ensure a more conceptual treatment of energy in the experimental teaching sequence. Energy stores can be filled or emptied along ways or pathways (Lawrence, 2007). Pathways provide information about the process or mechanism involved when energy is transferred from one store to another and also about the quantity of energy being transferred. The pathways are represented with an arrow notation. The thickness of the arrow represents the quantity of energy transferred along a pathway. Changes observed in a system as a physical process progresses, are described in terms of energy changes. The representation used presents the emptying of an energy store along with the simultaneous filling of other stores, with energy being transferred along a specific pathway. As energy is transferred from one store to others, the amount remains the same.

CYPRUS NATIONAL CURRICULUM FOR ENERGY FOR UPPER SECONDARY SCHOOL, FIRST GRADE

In the Cyprus Educational System, eight 45-minute lessons are allocated for teaching about energy to students in the first grade of upper secondary school (ages 15–16 years) (http://www.moec.gov.cy). The conceptual sequence appearing to underpin the Cyprus physics curriculum is characteristic of a traditional approach to teaching about energy with its origins in classical mechanics. The starting point of the sequence is the introduction of the concept of work, proceeding on to the introduction of the mechanical forms of energy (i.e. kinetic, gravitational, elastic) and ending with the introduction of the conservation of mechanical energy principle. Emphasis is placed on the quantitative analysis of energy problems, rather than an underlying conceptual understanding. Mechanical forms of energy are introduced independently with the transformation aspect dominating. The aspect of conservation is not explicitly introduced; rather it is introduced through the conservation of mechanical energy principle. This principle, however, is a conditional expression of the general principle of the conservation of energy:

it is valid only in the case of physical processes in which no energy degradation takes place. Moreover, the aspect of degradation is not explicitly introduced. Both the conservation and the degradation of energy aspects are discussed through a limited number of worked examples in which physical processes in systems take place under real conditions.

DESIGN OF THE TEACHING INTERVENTION

A critical review of approaches to design in science education was undertaken. From these, the perspective set out by Leach and Scott (2002) drawing on the concept of Learning Demand (Leach & Scott; Scott, Leach, Hind & Lewis, 2006) and the Communicative Approach (Mortimer & Scott, 2003) was selected. The learning demand design tool helps design decisions about content-specific teaching goals at a fine grain size. In the design process, Leach and Scott state "The purpose of identifying learning demands is to bring into focus the intellectual challenges facing learners as they address a particular aspect of school science; teaching can then be designed to focus on those learning demands" (p. 126). "Teaching goals" (Scott et al., p. 65) are then constructed to address learning demands. As Scott et al. (p. 76) stated: "Through this kind of analysis the 'conceptual terrain' of the teaching and learning domain is laid out and the teacher becomes more aware of what is involved for the students in learning in this area and what the teacher might do to support that learning."

Later in the chapter, we will illustrate some key differences in the pre-instructional views of learners, and the views of the physics curriculum for energy for students aged 15–16. We will show how these learning demands were used to inform design decisions in the development of the teaching sequence.

The purpose of the communicative approach design tool is to match explicitly the mode of teacher-to-whole-class talk to the teaching purpose. Classroom talk is characterized along two dimensions: interactive to non-interactive and dialogic to

	Interactive	Non-Interactive
Dialogic	A: Interactive/Dialogic: the teacher seeks to elicit and explore student's ideas about a particular issue with a series of 'genuine' questions.	B: Non-interactive/Dialogic: the teacher is at presentational mode (non-interactive), but explicitly considers and draws attention to different points of view (dialogic), possibly in providing a summary of earlier discussion.
Authoritative	C: Interactive/Authoritative: the teacher typically leads students through a sequence of instructional questions and answers with the aim of reaching one specific point of view.	D: Non-interactive/ Authoritative: the teacher presents a specific point of view.

FIGURE 9.1. Four classes of communicative approach

authoritative. Combining the two dimensions, four classes of communicative approach were identified: "interactive and dialogic," "non-interactive and dialogic," "interactive and authoritative" and, "non-interactive and authoritative." The four classes of communicative approach are explicitly described in Figure 9.1 (Scott et al., 2006, p. 73).

The framework set out by Leach, Scott and associates proposes a four-step scheme for the design of teaching sequences:

1. Identify the *school science* knowledge to be taught, drawing upon curriculum materials and other evidence as available and appropriate.
2. Drawing upon empirical evidence in the research literature, and other evidence as available and appropriate, consider students' likely interpretations for this part of the curriculum, before teaching begins.
3. Identify the *learning demands* by appraising the nature of any differences (concepts and associated ontology and epistemology) between 1) and 2).
4. Design a *teaching intervention* to address each aspect of this learning demand:
   a. Identify *teaching goals* for each phase of the intervention
   b. Plan a sequence of *activities* to address the specific teaching goals
   c. Specify how these teaching activities might be staged through appropriate *forms of classroom communication* using the communicative approach design tool.

In the following, an overview of the work done at each step of the design of the research-informed teaching sequence for energy is presented.

Step 1: The School Science Knowledge to be Taught

The first design decision was to develop, at a fine grain size, a theoretical framework about energy as consistent as possible with that used by physicists, that would be understandable by students, and that would meet the requirements of the Cyprus physics curriculum as previously described. As already indicated, we decided to draw upon Feynman's abstract perspective on energy. This perspective is well expressed for educational purposes within the SPT11-14 project (Institute of Physics, 2006) and we, therefore, used the SPT project as a basis for the development of the theoretical framework.

According to the proposed theoretical framework:

Energy is a unifying concept, intended to account for as wide a range of phenomena as possible (an epistemological feature). In terms of its ontology, it is abstract and mathematical in nature. *Energy is a property of systems* which is related to four key ontological aspects.

Aspect 1: Energy Can Be Stored

An amount of energy can be found in *energy stores*. The amount of energy in an energy store can be calculated in different ways for different stores. The term, energy store, is used to characterize the state of a system in a particular time interval of a process occurring in the system. We define four energy stores.

1. Gravitational store: when an object is at a raised position above the center of a planet or some other reference point, there is energy in a gravitational store of the system object-planet. The amount of energy of a gravitational store changes when the raised position of the object above the center of the planet or other reference point increases or decreases.
2. Kinetic store: a moving object has energy in a kinetic store. The amount of energy of a kinetic store changes when the magnitude of the velocity of the object increases or decreases.
3. Elastic store: a deformed (stretched or compressed) elastic object has energy in an elastic store. The amount of energy in an elastic store changes when the deformation (stretching or compression) of the elastic object increases or decreases.
4. Internal store: a solid, a liquid or a gas has energy in an internal store. Internal stores consist of two parts: the energy associated with the continuous motion of atoms/molecules/ions (kinetic store) from which an object or a substance is made up; and the energy associated with the chemical bonds between the particles themselves. The amount of energy of an internal store changes when the temperature of the object or the substance increases or decreases because the energy associated with the kinetic store increases or decreases.

We are aware that the use of the term energy store is an inclination from Feynman's theoretical framework for energy in which the term forms of energy is used. The forms of energy language are also commonly used by physics experts and physics educators. Our decision to use energy store instead of form of energy considered objections raised in the science education literature. Falk, Herrmann, and Schmid (1983) claimed the term energy form '.'.. is unsatisfactory because it easily leads to the interpretation that there are different kinds of energy, rather than emphasizing the simpler and physically more correct picture of energy as an unalterable substance" (p. 1074). A second objection was raised by Summers (1983) and Mac and Young (1987), who stated the use of forms of energy entails the risk that energy is considered as something contained in objects rather as a numerical quantity. An objection of another kind, concerning the emphasis placed on forms of energy was raised by Ellse (1988) who claimed "It draws attention away from the easier, more useful and important understanding of *energy transfer*" (p. 427). Furthermore, Ellse claimed the forms of energy approach deals with the form in which energy is manifested at different points rather than with the processes with which it is transferred from one place of a system to another or from one system

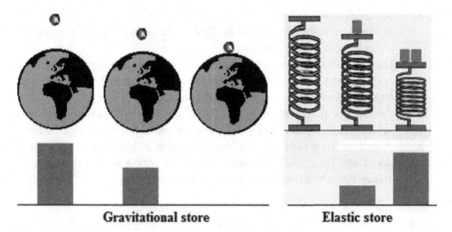

| Gravitational store | Elastic store |

FIGURE 9.2.   Representations of selected energy stores

to another. Furthermore, Millar (2005) also raises concerns about the use of the forms terminology. His first concern against the term is that students learn a set of labels which does not add much to their understanding. A second concern raised by Millar is that forms "lead to analyses of situations which introduce unnecessary variables that do not contribute to understanding" (p. 8). Moreover, he claims that, in some cases, the use of forms of energy lead to incorrect analyses of the process under study.

In our design the amount of energy in the energy stores was represented by a simple bar notation. Changes in the amount of energy in an energy store were represented with an increase or a decrease of the height of the bar. This particular abstract notation broadly used in graphical notations in economics, mathematics and statistics, was considered to be the most suitable to emphasize the abstract nature of energy, while avoiding the risk of considering energy as a kind of material. The representations presented in Figure 9.2 demonstrate the approach to visualizing energy stores used in the design. For the gravitational store, an object is presented falling to the Earth's surface, which in this example is considered as the reference point, with the bar chart showing the amount of energy in the gravitational store. In the case of the elastic store, a spring is compressed by similar masses and the bar chart shows the amount of energy in the elastic store of each state of the compressed spring.

Aspect 2: Energy Can be Transferred Along Transfer Pathways from One Energy Store to Another

Transfer pathways describe the process or the mechanism involved during the transfer of energy. We define three transfer pathways.

1. Mechanical working: when a force is exerted and its point of application is displaced in the direction of the force, then energy is transferred along a mechanical working pathway.
2. Heating: when two objects or substances of different temperature are in contact, then energy is transferred along a heating pathway. Energy is transferred either by the process of conduction primarily in solids, by the process of convection in liquids and gases or by radiation, a process which can take place both in the presence and in the absence of matter.
3. Sound: when a cause (for example, the clap of hands, the knock on a door) sets the particles (atoms, molecules) within an object or a substance into vibration, then energy is transferred along a sound pathway.

Transfer pathways were represented with the simple notation of an arrow. The direction of the arrow defines the direction of the energy transfer whereas the thickness of the arrow represents the amount of energy being transferred from one energy store to another.

Aspect 3: Energy is Conserved

As the amount of energy of a store is decreased, the amount of energy of some other stores is increased and thus, the quantity of energy of a system remains unchangeable.

Aspect 4: Energy is degraded.

It is always the case that energy transfers result in some energy being transferred into an energy store which cannot then be transferred along another transfer pathway.

Changes occurring in a system as a physical process can be described in terms of energy changes. An account of the change process in energy terms will talk about the emptying of energy stores and associated filling of other energy stores, with energy being transferred along specific pathways. As energy is transferred from one store to others, the amount remains the same. We are aware that the language of filling and emptying can be interpreted in a materialist way: the challenge for learners was to become able to use the new terminology to account for changes in phenomena in terms of energy without resorting to materialistic models.

Describing Changes in Terms of Energy

In order to describe changes in terms of energy, it is necessary to define the physical system in which the changes are taking place. Within this theoretical framework, the group of objects involved in the changes is called a *system*.

The transfer of energy within a system was represented diagrammatically both qualitatively and quantitatively with a *Sankey diagram* (Hobson, 2004; Millar, 2005; Scherr et al., 2012). In a Sankey diagram, the different kinds of energy stores are represented with a simple rectangular bar, whereas changes in energy stores are represented with the increase or the decrease of the height of the bar. Also, the different kinds of transfer pathways are represented by an arrow. In a Sankey diagram, however, only the transfer of energy from an initial to the final energy store(s) can be represented. For a detailed representation of the energy transfer of a system from the initial to the final store(s), it is proposed that the course of energy be represented with a *Full Sequence Energy Diagram* (Orfanidou, 2014). In a Full Sequence Energy Diagram, each energy transfer from one energy store to another is represented as a single sub-event. Thus, the course of energy from the initial to the final energy store or stores is represented as a sequence of different sub-events.

The physical changes occurring in a system were represented through a sequence of four different descriptions: 1) the problem statement, a description serving as a source of information for both understanding the process taking place in a system and for providing quantitative data; 2) the physical description, a verbal description of the changes observed in a system in the natural world; 3) the energy description, a verbal and diagrammatic description of the changes observed in a system, expressed in terms of energy and, 4) the mathematical description, a quantitative description of both the changes in the amount of energy stores and the amount of energy being transferred along the transfer pathways, as a physical process takes place in a system. Although this design decision was taken based upon the professional insight of the first author based upon her physics teaching experience, it is also consistent with Tiberghien and associates' "two worlds" framework described by Ruthven, Laborde, Leach, and Tiberghien (2009, p. 335). The framework requires students to broaden their language to encompass the world of objects and events and the world of theories and models, and between physics and everyday explanations in both of these worlds. An example in which the sequence of the descriptions is applied for the analysis of the changes between two snapshots of a process is presented in Figure 9.3.

The theoretical framework presented shares features and differences compared to theoretical frameworks of approaches developed by other researchers. For example, Van Heuvelen and Zou (2001) proposed the use of multiple representations for the study of work-energy processes by students attending college introductory physics courses. According to this approach, a work-energy process can be described through four kinds of representations: the verbal, the pictorial, the bar chart and the mathematical representation. A verbal representation of a process is a description of the process in words. A pictorial representation of a process is a sketch or a picture which represents the process. A physical representation involves the construction of an energy bar chart to describe work-energy process. Finally, a mathematical description involves the use of the mathematical formula of the work-energy theorem to describe the process quantitatively.

We share the view expressed by Van Heuvelen and Zou that the use of multiple representations for the study of physical processes improves student achievement

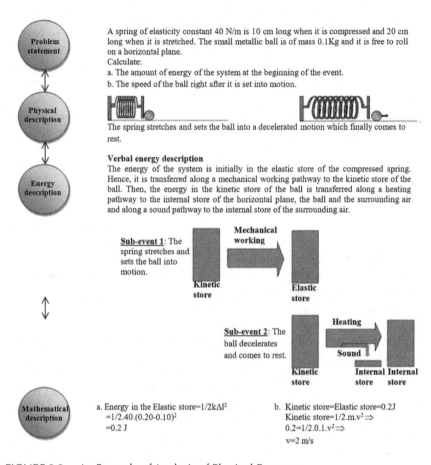

Problem statement

A spring of elasticity constant 40 N/m is 10 cm long when it is compressed and 20 cm long when it is stretched. The small metallic ball is of mass 0.1Kg and it is free to roll on a horizontal plane.
Calculate:
a. The amount of energy of the system at the beginning of the event.
b. The speed of the ball right after it is set into motion.

Physical description

The spring stretches and sets the ball into a decelerated motion which finally comes to rest.

Energy description

Verbal energy description
The energy of the system is initially in the elastic store of the compressed spring. Hence, it is transferred along a mechanical working pathway to the kinetic store of the ball. Then, the energy in the kinetic store of the ball is transferred along a heating pathway to the internal store of the horizontal plane, the ball and the surrounding air and along a sound pathway to the internal store of the surrounding air.

Sub-event 1: The spring stretches and sets the ball into motion.

Kinetic store — Mechanical working — Elastic store

Sub-event 2: The ball decelerates and comes to rest.

Kinetic store — Heating / Sound — Internal store / Internal store

Mathematical description

a. Energy in the Elastic store=$1/2k\Delta l^2$
=$1/2.40.(0.20-0.10)^2$
=0.2 J

b. Kinetic store=Elastic store=0.2J
Kinetic store=$1/2.m.v^2 \Rightarrow$
$0.2=1/2.0.1.v^2 \Rightarrow$
v=2 m/s

FIGURE 9.3. An Example of Analysis of Physical Processes

in problem-solving. In our view, this is particularly true for enhancing conceptual understanding of the abstract concept of energy by students and for providing them an explicit methodological scheme for the description of physical processes and events occurring in a system in terms of energy. The two approaches share similar corresponding descriptions such as the verbal and the problem statement, and the mathematical description. Pictorial representation in the multiple representations approach involves a sketch or a picture which depicts the system in which the work-energy process takes place. Rather, in our approach, the corresponding physical representation includes a picture or a computer simulation presenting a system in which an event or a process takes place alongside a verbal description of the event or process. Our decision to include the verbal description of the pictorial representation of an event or a process was based on the first author's observations during teaching practice that many students could not recognize all aspects of the event or process they were observing. As a result, they could not describe fully and correctly the event or process in terms of energy. Last, the bar chart representation in the multiple representations approach includes the construction of energy bars in energy-initial and final state of the system axes. Our corresponding energy' representation includes a detailed verbal and diagrammatic description of the course of energy from the initial to the final state of the system in which a process occurs. We believe students could represent diagrammatically more easily and correctly the course of energy of a process if they could first identify all energy stores and pathways through which energy is transferred and use them fully to describe the event or process in terms of energy both qualitatively and quantitatively.

Another approach which shares features but also the key difference concerning the ontology of energy with our research-informed teaching approach for energy is the substance metaphor of energy proposed by Brewe (2011). According to this metaphor, energy is considered as a property of systems which can: 1) be stored in both objects and interactions, 2) be transferred from one storage location to other(s) causing changes within a physical system and, 3) be conserved. These comprise key elements of our research-informed teaching approach for energy. There are, however, significant differences concerning the conceptualization of these elements due to the different ontological consideration of energy between the two approaches. First, in our approach the aspect of storage is defined through an abstract perspective. The term energy store is used for the characterization of the state of a system in a particular time interval of an event or process occurring in the system. In the substance metaphor approach, a kind of substance, namely energy, is stored in the objects which comprise a system or the interactions which take place in it. Second, the aspect of transfer in our approach is used to characterize the mechanism or process which takes place for the changes in the state which a system undergoes within a particular time interval of an event or a process. In both approaches energy is considered to transfer from one storage place to other(s). Furthermore, in the substance metaphor approach, the transfer is associated with the flow of some kind of a substance, namely energy. Most importantly, the energy transfer is considered as a causal agent of the changes occurring in a system, which in our view is an inclination of the scientific view of energy. Third, in our approach the aspect of conservation refers to a number which should be the same before and after an event or a process in a system. In the substance metaphor approach, conservation refers to a quantity of a substance, namely energy, which remains unchangeable before and after the conclusion of a process.

Another common feature of the two approaches is the emphasis placed on the concept of a system. According to Brewe: "Two ways that scientists regularly utilize energy is to identify the relevant objects and interactions present in a physical system and to identify energy storages in the application of energy conservation" (p. 020106-6). We share the view that students should be able to identify the objects which comprise the physical system under study in order to understand conservation and we add to this, in order to understand degradation as well. This view is supported by Daane, Vokos, and Scherr (2014) who state "Energy degradation depends on a specific system (which comprises all relevant objects of a scenario),..." (pp .020111–2). The importance of identifying all parts of a system is highlighted in the multiple representations of work-energy processes approach (Van Heuvelen & Zou, 2001).

The importance of the use of visual representations to promote conceptual understanding of energy is another shared feature between Brewe's substance metaphor of energy and our research-informed teaching sequence for energy. In the substance metaphor approach, the use of energy pie charts and energy bar charts is suggested. In our approach, we developed the Full Sequence Energy Diagram, which is a combination of a Sankey diagram usually comprising two or more branches and energy bars. As Brewe remarks, pie and bar energy charts can represent qualitatively and quantitatively the energy stores involved in a physical event or a process and enhance understanding of conservation. In our view, the Full Sequence Energy Diagram goes further. It provides the students with a detailed qualitative and quantitative picture of the course of the energy of a system by visualizing all energy stores and transfers involved in a quantitative manner. This kind of visual representation emphasizes the aspects of conservation and degradation helping students in the next step, namely, the mathematical description of a process in terms of energy.

A third approach which shares features and differences compared to our research-informed teaching approach for energy is the substance metaphor proposed by Scherr, Close, McKagan, and Vokos (2012). According to the researchers, the substance metaphor supports the following features: 1) energy is conserved; 2) energy is localized; 3) energy is located in objects; 4) energy can change form; 5) a) energy is transferred among objects and b) energy can accumulate in objects. This substance metaphor is based on a similar theoretical framework with substance metaphor proposed by Brewe (2011) already noted in previous paragraphs. A difference between the two substance metaphors is that in the Scherr et al. metaphor (and in our research-informed teaching sequence for energy), energy is considered to be transferred from an energy storage to other(s) whereas in Brewe's substance metaphor, energy is transferred between objects. Moreover, Scherr et al. state the consideration of energy as being located in objects is a limitation of their approach for gravitational and other forms of potential energy, which are well defined within a system of objects or in a field, rather in individual objects. For an abstract consideration of energy, there is not such a concern and thus there is no such limitation.

Step 2: The Students' Likely Interpretations about Energy

There is a wealth of material in the international science education literature investigating how students at the ages of 15 and 16, and at the concomitant educational stage, interpret phenomena and events featured in the curriculum outlined above. Gilbert and Pope (1986) proposed seven conceptual models, formulated based on the findings available in the literature at that time, typically used by students to interpret phenomena and events in this part of the physical science curriculum: a) anthropocentric— energy is associated with living organisms and specifically with human beings and also with objects considered to possess human characteristics (Black & Solomon, 1983; Solomon, 1983; Stead, 1980; Watts & Gilbert, 1985); b) depository— specific substances, objects or media such as fuels, food and batteries can store energy, need energy or consume energy which is stored in them. (Ault, Novack, & 1988; Gilbert & Pope, 1982; Solomon, 1983; Watts & Gilbert); c) ingredient—energy is associated with fluids or ingredients that are dormant and are released suddenly by a trigger (Watts & Gilbert); d) activity: energy is associated with motion, force and activity (Brook & Driver, 1984; Duit, 1981; Gilbert & Pope, 1982; Watts & Gilbert, 1983); e) product—energy is viewed as a kind of by-product of a situation that is generated, is active, and then disappears or fades (Watts & Gilbert, 1985); f) functional—energy is perceived as a fuel the amounts of which are limited (Ault et al.; Duit; Stead, 1980); and g) flow-transfer—energy is considered as a fluid which can flow from one object to another (Gayford, 1986; Duit; Stead; Watts & Gilbert, 1985).

The conservation of energy is a fundamental idea related to the nature of the concept of energy. Research, however, suggests there is a poor scientific understanding of the idea (Ault et al., 1988; Boyes & Stanistreet, 1990; Duit, 1981; Driver & Warrington, 1985). In order to check that Cypriot students' interpretations of phenomena and events prior to studying this part of the curriculum were broadly consistent with what had been reported in the international research literature, diagnostic questions were administered to groups of students and analyzed. We found the interpretations students used were consistent with Gilbert and Pope's (1986) framework. Of the seven conceptual models identified, the most frequently used were activity, depository and anthropocentric. We additionally found Cypriot students did not appreciate the law of conservation of energy often talking about energy as a thing that can "disappear," "stop existing" or "be consumed."

Step 3: Identify the Learning Demands

Learning demands were identified by comparing the differences between the framework for energy described in step 1, and the interpretations of phenomena and events forming the focus of this part of the physics curriculum before instruction as described in step 2. We now present an example of one learning demand arising from comparisons between the views about energy within the content of school science and the pre-instructional descriptions typically used by Cypriot physics learners prior to studying this part of the curriculum. This is an illustrative example; other learning demands were identified leading to teaching goals and design decisions about teaching activities. In step 1, we noted the curriculum requires students to be introduced to a concept of energy where *the total amount of the energy of a system remains constant*, and yet, in step 2 we saw Cyprus students often talk about energy as a thing that can disappear, stop existing or be consumed. Thus, there is a fundamental ontological difference between how the word energy is used in the context of school science, and the views of Cypriot physics students prior to studying this part of the curriculum. We, therefore, identified a learning demand for students as coming to see energy as a conserved quantity rather than something that can disappear or be consumed or stop existing at the end of a physical process. The teaching goal identified from this learning demand was *students should be enabled to understand how the energy of a system is conserved through change processes.*

Step 4: Design a Teaching Intervention to Address each Aspect of the Learning Demand

The next set of design decisions involved developing activities to meet the teaching goals identified, and deciding how those activities were to be staged in the classroom through teacher talk. We used the communicative approach design tool to inform decisions about teacher talk. We present how the above teaching goal was addressed through the design of a teaching activity at a fine grain size, and how it was decided to stage (Leach & Scott, 2002) that activity in the classroom through teacher talk using the communicative approach design tool.

The activity was based on a computer simulation, *Sim 3*, we developed to serve the above teaching purpose on conservation of energy, and was intended to take about 15 minutes. The activity was designed to provide students with the opportunity to practice using the new ideas introduced in the first lesson, that is, to account for an event in terms of the new ontological entities *energy store* and *transfer pathway* within a broad conceptual framework. The teacher presents Page 1 of *Sim 3* and asks students to study it carefully. Page 1 of *Sim 3* presents a smooth, semi-spherical vessel in which a small ball rolls from its upper point (A) to the bottom (B) and then to the other upper point (C) without friction. The ball's movement in the vessel is illustrated in the screenshots in Figure 9.4.

The example was designed to challenge students to explain the behavior of an event in terms of the new energy ideas. In particular, the points of zero velocity (A) and (C) require students to confront a key difference between the pre-instructional view that there is no energy, and the newly introduced view that the energy of the system is all in the gravitational store.

The teacher then asks the students to recall the key elements of the theoretical framework for energy introduced in Lesson 1 through an interactive and authoritative approach by posing the following questions: "Describe the changes which you observed to occur in the simulation. Can you use the energy concept to interpret in detail these changes?" As the teaching purpose was to support students in using the energy ideas as introduced in the first lesson, an authoritative approach was used.

Next, the teacher presents Page 2 of simulation *Sim3* asking students to study it carefully. Page 2 presents the ball going through frictionless motion from the upper

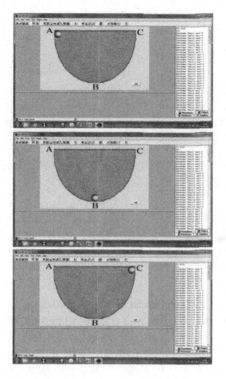

FIGURE 9.4.   Screenshots of Page 1 of Simulation *Sim 3*

point A of the vessel to the other upper point C again and again. The teacher invites
students to account for what they are seeing in terms of energy using an interactive
and dialogic approach. The teaching purpose is to enable the teacher to ascertain
whether the students spontaneously draw upon the conservation of energy in their
accounts. The following question was posed: "As we see, the ball repeats exactly
the same motion. As we also notice, the ball stops instantaneously at points A and
C. Is there any amount of energy stored in the system at these points?" (Quotations
have been translated from the original Greek by the first author.). It was expected
that some of the students would construct their answers drawing upon their pre-
instructional views, stating there is no energy stored in the system at points A and C.

Then, the teacher undertook further probing and posed the following question:
"OK. If there is no energy stored in the system at points A and C, how can you
interpret the fact that the ball repeats its motion over and over?" By showing the
ball in constant motion, the simulation forced students to confront differences
between their pre-instructional views and the newly introduced energy views, and
provided opportunities for them to practice using the ontological entities *energy
store* and *transfer pathway*. A dialogic approach was used in order to enable the

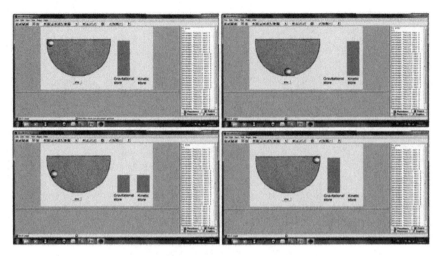

FIGURE 9.5. Screenshots of Page 3 of Simulation Sim 3

teacher to explore the students' understanding and diagnose the most appropriate action in order to achieve the teaching goal of supporting the students in becoming able to use the energy ideas introduced in the first lesson.

Next, the teacher reviewed the students' ideas through a non-interactive and dialogic approach. The following stage of the lesson involved the teacher presenting Page 3 of *Sim3* and asking the students study it carefully again. In Page 3, the amount of energy in the gravitational store of the system ball (Earth) and in the kinetic store of the ball are shown, as the ball moves from the upper point A of the vessel to the lowest point B and then to the other upper point C. A few screenshots of the simulation are illustrated in Figure 9.5.

Through an interactive and authoritative approach, the teacher moves towards using the newly-introduced ideas about energy stores and transfer pathways reinforced by the on-screen graphics. The following question was posed: "What do you conclude about the total amount of energy stored in the system?" An interactive and dialogic approach enabled the teacher to assess the students' ability to use the newly introduced energy ideas to construct an interpretation of the presented phenomenon. The teacher, however, moved into a non-interactive and authoritative approach to illustrate how to interpret the simulation in terms of the new energy ideas, with a particular emphasis on energy conservation. Finally, a definition of the conservation of energy was introduced.

IMPLEMENTATION AND EVALUATION
OF THE TEACHING SEQUENCE

This section addresses the implementation and evaluation of the teaching sequence around three research questions:

Research Question 1: Are the concepts of energy used by a Cypriot cohort of upper high school students prior to teaching consistent with those reported in the international literature, and consistent between experimental and comparison groups?

Research Question 2: Was the research-informed teaching sequence successful in promoting enhanced conceptual understanding about energy when compared with the outcomes of students following the usual instructional approach?

Research Question 3: How effective is the research-informed teaching sequence in promoting conceptual understanding with students of different prior attainment in physics?

DESIGN AND METHOD OF THE STUDY IN OVERVIEW

The research was conducted in a real educational setting following a quasi-experimental design (Campbell & Stanley, 1963). A *pre-test-post-test non-equivalent control group design* (Wiersma, 1986) was used to address the first two research questions. This approach allowed us to implement the designed teaching sequence with one group of students and have a comparison group who were taught the same content using the school's usual approach.

About one month prior to the start of teaching, the teacher of the experimental group was provided with a full set of the teaching materials. About a week before the start of teaching, the teacher met with the first author to raise any questions she had about the implementation of the teaching. In implementing the teaching, the teacher reported attempting to be as faithful to the design as possible.

The quasi experiment was carried out in three stages. During the first stage, the pre-instructional interpretations about energy of the students participating in the experimental and comparison groups were collected through a pre-test. In the second stage, the experimental intervention and further data collection through individual interviews with a small cohort of experimental students and the class teacher occurred. Interview data were used to address the third research question. Finally, in the third stage, the students' interpretations about energy in both the experimental and comparison groups after the intervention were collected through a post test.

The teaching in the experimental group occurred in eight 45-minute lessons as defined by the Cyprus National Curriculum. In staging the energy knowledge, the introduction and qualitative treatment of the proposed energy model took place first and were followed by the quantitative treatment. The first three of the 8-lesson teaching sequence included the introduction of all four aspects of the energy concept (i.e. store, transfer, conservation and degradation) and their use in describing both verbally and diagrammatically the changes occurring in various physical systems. In the remaining five lessons the specific energy stores defined by the curriculum, namely, kinetic, gravitational, elastic and mechanical store, and the specific transfer pathways of mechanical, heating and sound, were further elaborated qualitatively and mathematically.

In addition to staging the energy knowledge, the students' internalization of the specific ideas introduced was supported by providing opportunities through which students tried out the new ideas during discussions both with the teacher and between themselves. In addition, short written tasks were set to be undertaken during the lessons either by individual students or in groups.

The teaching in the comparison group took the form of an eight lesson sequence as would usually take place in the school. The energy knowledge was staged according to the conceptual sequence defined by the National Curriculum presented earlier in this chapter. It should be noted that normal teaching involves the teacher in introducing energy ideas through an authoritative traditional approach (Mortimer & Scott, 2003) and by following the book. Student internalization of the specific ideas introduced was supported through short written tasks included in the book to be done during the lessons by individual students, and reviewed and corrected on the whiteboard by a student or the teacher.

A video recording was made of all lessons among both the experimental and comparison groups. The recordings were used to assess the content and conduct of the teaching after it had taken place, should we have any questions about the treatment of content or use of teacher-to-whole-class talk in either group.

Participants

The research study was conducted in an urban public upper secondary school in Limassol, Cyprus. Of the five full-time physics teachers of the school, two volunteered to act as either the experimental or the comparison teacher. The two teachers had similar professional characteristics: they were physics experts, trained teachers, had similar instructional technology (IT) skills, had the same number of years in service, and the same number of years teaching students of this age. The teachers themselves decided which of them would teach the experimental and comparison groups.

Each teacher selected one of the three classes of students of the target age in which they were teaching physics to participate in the study. The two classes were of mixed ability. The main criterion for their selection was the relationship between the students and the teacher. Of the 42 students studying in the two classes, 36 consented to participate in the research, 18 students in each of the experimental and the comparison group respectively. As well as comparing students' responses to pre-test questions, we also looked at data on their prior attainment in physics held by the school to establish there were no obvious differences in prior attainment between the groups.

Data Collection

Pre-test

Pre-test data were collected for two purposes. To address the first research question it was necessary to establish the kinds of reasoning used by students in the sample prior to teaching, to describe phenomena and events accounted for using energy concepts in physics. Secondly, to assess the effectiveness of the designed teaching compared with the school's usual approach (research question 2), it was necessary to establish comparable understandings of energy among the experimental and comparison groups. A phenomenological approach (Driver & Erickson, 1983) was used. The approach involves presenting students with an event or a system and, through a set of diagnostic questions, allowing students to interpret the behavior of the event or system in whatever terms they wish to. Data from the pre-test could then be used to characterize the reasoning used by students in the sample before teaching, to compare their responses with those documented in the international research literature, and to compare them with the students' responses after teaching using either the designed teaching sequence or the school's usual approach.

The pre-test consisted of two parts. Part A included diagnostic questions addressing areas addressed through the Cyprus physics curriculum with students of a younger age (to enable us to judge the comparability of the two teaching groups). The set of questions referred to the study of a physical process taking place in a system presented through a simulation. Part B included diagnostic questions to see which models might be used by students to generate interpretations, in order to compare these with those identified in the research literature.

Three expert reviewers reviewed the pre-test. Each expert reviewer held a first degree in physics, a doctoral degree in science education and was an active researcher in physics education. The experts were encouraged to suggest possible changes concerning the physical systems included the instruments and the formulation of the tasks in them. This review procedure provided an indication of the content validity of the data collection instruments (Cohen, Manion, & Morisson, 2007). The revised version of the pre-test was pilot tested with one first grade secondary school class, taught by a physics teacher who was not otherwise involved with the study. The pilot testing provided evidence for the appropriateness of all data collection instruments since their development was based on a similar rationale and similar wording was used. No indication of serious misunderstanding of any of the tasks emerged and the data collection instruments underwent minor further amendments.

Students' responses to both parts of the pre-test were analyzed ideographically (Driver & Erickson, 1983): students' responses were coded in their own terms rather than being assessed against correct physics knowledge. For part A, students' responses were reviewed and grouped together. Through a process of refinement, coding categories were identified and frequency counts of students' responses were constructed for each category. Differences in frequency counts between the experimental and comparison groups were analyzed using Fischer's Exact Test (Field, 2009). Students' responses to the diagnostic questions in part B were initially treated in a similar way, and coding categories were developed. These categories were then compared to the models identified by Gilbert and Pope (1986). Categories were also included for responses using a correct energy model, or no implicit model. Differences between the frequency of responses between the experimental and comparison groups were also compared using Fischer's Exact Test. Analysis of data was conducted by the first author. For the investigation of the inter-rater reliability of the results, a selection of students' responses, randomly selected from both groups, was independently coded by the three expert reviewers of the data collection instruments. No serious disagreements were detected between the first author and the reviewers' interpretations of the students' responses and minor disagreements were easily resolved. We concluded the coding scheme was communicable and valid. The pre-test questions are presented in Appendix A.

Post-test

The post-test aimed to collect data which would provide information concerning the students' interpretations about energy after the teaching interventions. The post-test was structured in two parts. Part A was the same as the corresponding part A of the pre-test. The questions in part B were conceptually framed (Driver & Erickson, 1983), presenting students with words or concepts from physics and asking specific tasks be performed with them. This process enabled us to judge the extent to which the students were able to use energy concepts introduced during the teaching. Students were presented with two novel physical systems, illustrated in two images, and were asked to produce energy descriptions of the systems in both qualitative and quantitative terms. The systems had not been used in teaching. We were able to assess the extent to which students could use the introduced energy ideas to produce energy descriptions of unfamiliar systems. Post-test questions were checked for reliability, validity and ease of implementation through a similar process to that described for the pre-test questions.

Post-test part A questions were analyzed using the same approach as for pre-test part A. This allowed a direct comparison between students' pre- and post-teaching responses in both the experimental and comparison groups. In addition, the post-test part A questions were analyzed nomothetically (Driver & Erickson, 1983). Such analysis enabled us to comment upon the extent to which the students produced energy descriptions (both qualitative and quantitative) consistent with the aims of the teaching.

Part B questions were also analyzed nomothetically. Students' responses were classified into groups according to whether their energy descriptions were correct, partially correct or incorrect and frequency counts were constructed and compared. The post-test questions are presented in Appendix B.

Student Interviews

We collected interview data from a sub-set of four students from the experimental group. The students were selected to include two girls and two boys, with two students of previous high attainment in physics and two of previous low attainment. Two interviews were conducted, one during the teaching at a point when the four key aspects of energy had been introduced (i.e. store, transfer, conservation, degradation), and another towards the end of teaching. These interviews aimed to collect data providing deeper and more detailed insights concerning the students' understanding of the energy ideas taught through the research informed teaching sequence. The interviews were semi-structured (Wilkinson & Birmingham, 2003) with the open-ended key questions included in the interview protocols (Anderson & Arsenault, 1998). We considered this number of students to be interviewed as sufficient for illustrative purposes, given that the sample comprised more than 10% of the participants.

The set of key questions in each of the interview protocols referred to a physical process taking place in a system illustrated in an image. The key questions were grouped in one of two thematic sub-sets. The first sub-set probed the students' ability to describe the physical process in terms of energy. Students were provided with two Full Sequence Energy Diagrams. The second sub-set probed the students' understanding of the energy ideas introduced (i.e. store, transfer, conservation, degradation). We drew upon data from these interviews, together with data from the students' responses to the pre- and post-tests, to illustrate how their conceptual understanding changed as a result of following the teaching as well as their affective response to the teaching. The interview protocol of the first and the second interview is presented in Appendix C and D respectively.

Teacher Interviews

The interview with the experimental group teacher focused upon her opinions about the effectiveness of the designed teaching sequence in promoting conceptual understanding among students across a spectrum of prior attainment in physics, and the extent to which it was practical to use and enjoyable for her and her students. The interview was semi-structured with a protocol comprised of a set of open ended key questions grouped in two sub-sets. The first sub-set probed, in detail, the teacher's reactions to each lesson of the teaching sequence, addressing both the treatment of content and the use of teacher-to-whole-class talk. The second sub-set addressed the teacher's overall evaluation of the teaching sequence, including the number of lessons in the teaching sequence, its structure and the order in which the energy ideas were introduced. We drew upon this interview data to inform our judgements about the third research question.

THE EVALUATION OF THE RESEARCH INFORMED TEACHING SEQUENCE FOR ENERGY

In this section, we draw upon data from the data sets described above to support claims around the three research questions outlined earlier in this chapter.

Research Question 1

Pre-test part A was used to provide evidence about the knowledge about energy used by students in the experimental and comparison groups before teaching, to ensure any differences observed after teaching were not due to differences between the initial starting points of students. Students' responses were used to assess whether they used energy in their interpretations, or some other physics concept (particularly force); whether students interpreting in terms of energy did so in terms of the four aspects of energy in a manner consistent with the aims of the teaching; and whether any interpretations were used by students that have been previously identified in the international literature.

Only a very small number of the participant students used energy in their interpretations for the event portrayed in the simulation; instead, the vast majority used a force-based interpretation. These data are presented in Figure 9.6.

A statistical analysis of the above data revealed the p value for Fisher's Exact Test is $p=0.338>p_{cutedge}=0.05$ (Table 9.1), which suggests no significant difference in the number of students who used energy in their interpretations in the two groups.

For the four aspects of energy identified in the conceptual analysis these findings suggest that, prior to instruction for both the experimental and comparison groups: 1) there was some understanding of the energy store aspect, 2) there wasn't significant understanding of the energy transfer aspect, 3) there was a very weak understanding of the conservation of energy aspect and, 4) there wasn't significant understanding of the degradation of energy aspect. The relative absence of understanding of the participant students about the conservation of energy aspect prior to teaching is illustrated in the bar chart shown in Figure 9.7.

A statistical analysis of the data concerning the students' initial understanding of each of the four aspects of the energy concept suggested no significant difference between the two groups. In the case of the conservation of energy aspect, the p value for Fisher's Exact Test is equal to $p=1.000>p_{cutedge}=0.05$ (Table 9.2),

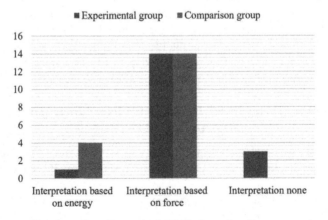

FIGURE 9.6. Number of students who initially formulated an energy based or a force based interpretation

TABLE 9.1. Results of Fisher's Exact Test About the Number of Students in the Two Groups Who Initially Formulated an Energy-Based Interpretation

	Value	df	Asymp. Sig. (2-sided)	Exact Sig. (2-sided)	Exact Sig. (1-sided)
Pearson Chi-Square	2.090[a]	1	.148		
Continuity Correction[b]	.929	1	.335		
Likelihood Ratio	2.218	1	.136		
Fisher's Exact Test				.338	.169
N of Valid Cases	36				

[a]2 cells (50.0%) have expected count less than 5. The minimum expected count is 2.50.
[b]Computed only for a 2×2 table.

providing evidence there is no significant difference between the number of students who stated the energy of the system is conserved between the two groups.

We used pre-test part B to investigate the different initial ideas used by students. Findings suggested students in both the experimental and the comparison group expressed ideas about energy which appear to be very similar to those reported in the international literature. The most frequently used is the activity model, followed by the depository and the anthropocentric models respectively (Gilbert & Pope, 1986). Moreover, the comparative study of the results of the experimental and the comparison group showed each of the most frequently used alternative models appeared with about the same frequency among students in the two groups. The findings for other parts of the pre-test part B could also be classified according to Gilbert and Pope's (1986) classification. We concluded the

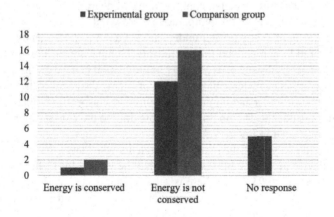

FIGURE 9.7. Number of students who initially stated that energy is/is not conserved

TABLE 9.2. Results of Fisher's Exact Test About the Number of Students in the Two Groups Who Stated Initially That Energy is Conserved.

	Value	df	Asymp. Sig. (2-sided)	Exact Sig. (2-sided)	Exact Sig. (1-sided)
Pearson Chi-Square	.364ª	1	.546		
Continuity Correctionᵇ	.000	1	1.000		
Likelihood Ratio	.370	1	.543		
Fisher's Exact Test				1.000	.500
N of Valid Cases	36				

ª2 cells (50.0%) have expected count less than 5. The minimum expected count is 1.50.
ᵇComputed only for a 2×2 table.

students in our sample used ideas about energy prior to teaching similar to those reported in the international science education literature.

Research Question 2

The post-test part A was used to assess the extent to which students could formulate an energy description of the event portrayed in the simulation after teaching, and the extent of students' understanding of the four aspects of energy targeted in the designed teaching (store, transfer, degradation, conservation). Most students in the experimental group tended to formulate an energy based description of the event portrayed in the simulation, compared to less than half of the students in the comparison group, as shown in Figure 9.8.

A statistical analysis of the number of students making energy based responses in the two groups (Table 9.3) shows that the p value for Fisher's Exact Test is $p=0.005<p_{cutedge}=0.05$, indicating a significant difference between the two groups.

Concerning the store, transfer, conservation and degradation aspects of the energy concept, findings from the comparative representation and the Fisher's Exact Test suggest a significantly greater understanding among experimental group students compared to comparison group students. Comparison of pre- and post-instructional stage results suggests a greater improvement in the experimental group students' performance in describing in terms of energy the event of the simulation, as well as in their understanding each of four energy aspects, compared to comparison group students. The findings corresponding to the conservation of energy aspect are illustrated in Figure 9.9.

A statistical analysis of the number of students who stated that energy is conserved in the two groups (Table 9.4) shows that the p value for Fisher's Exact Test is $p=0.075<p_{cutedge}=0.05$. Thus, there is a significant difference in the number of students who stated the energy of the system is conserved in the two groups.

Overall, the results suggest a significant difference in the experimental group students' understanding of the conservation aspect of energy compared to comparison group students. Comparison of pre- and post-instructional stage results

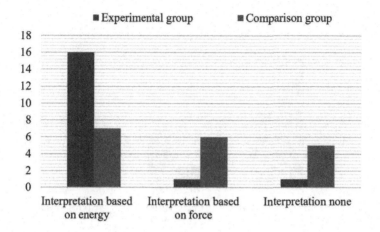

FIGURE 9.8. Number of students who formulated an energy based or a force based interpretation after instruction

revealed more improvement in the experimental group students' understanding of this aspect compared to comparison group students.

Post-test part B was designed to assess the extent to which students could produce energy descriptions of physical systems not encountered during teaching using the four aspects (store, transfer, conservation, and degradation). We draw two conclusions from students' responses to post-test part B. First, students from the experimental group were better able to use the taught energy ideas in novel physical systems than students in the comparison group, generating both qualitative and quantitative descriptions. Second, students' responses to the in the experimental group were better able to use knowledge of each of the four aspects of energy across the range of physical systems in both parts A and B of the post-test.

The findings summarized above provide evidence that the research-informed teaching sequence was successful in promoting enhanced conceptual understand-

TABLE 9.3. Results of Fisher's Exact Test About the Number of Students in the Two Groups Who Formulated An Energy-Based Interpretation After Instruction.

	Value	df	Asymp. Sig. (2-sided)	Exact Sig. (2-sided)	Exact Sig. (1-sided)
Pearson Chi-Square	9.753[a]	1	.002		
Continuity Correction[b]	7.706	1	.006		
Likelihood Ratio	10.477	1	.001		
Fisher's Exact Test				.005	.002
N of Valid Cases	36				

[a]0 cells (.0%) have expected count less than 5. The minimum expected count is 6.50.
[b]Computed only for a 2×2 table

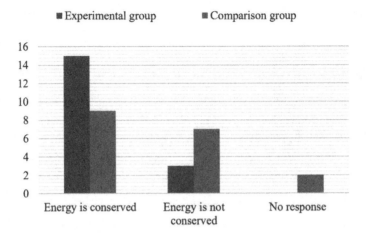

FIGURE 9.9. Number of Students Who Stated That Energy is/is Not Conserved After Instruction

ing about energy when compared with the outcomes of students following the usual instructional approach. Students who followed the research informed teaching sequence were better able to use the four aspects of energy targeted in the teaching (store, transfer, conservation, degradation) than their peers who followed the usual teaching approach. Furthermore, experimental students were better able than their peers to transfer their knowledge to construct energy descriptions of phenomena not encountered during teaching.

Research Question 3

We were conscious that our chosen conceptual approach had been rejected by other designers on the grounds it is too challenging for students. We were keen to

TABLE 9.4. Results of Fisher's Exact Test About the Number of Students in the Two Groups Who Stated That Energy Is Conserved After the Teaching Intervention

	Value	df	Asymp. Sig. (2-sided)	Exact Sig. (2-sided)	Exact Sig. (1-sided)
Pearson Chi-Square	4.500[a]	1	.034		
Continuity Correction[b]	3.125	1	.077		
Likelihood Ratio	4.656	1	.031		
Fisher's Exact Test				.075	.038
N of Valid Cases	36				

[a] 0 cells (.0%) have expected count less than 5. The minimum expected count is 6.00.
[b] Computed only for a 2×2 table

evaluate the outcomes of students of different previous attainment levels in physics as a result of following the designed teaching. Four case studies of individual students in the experimental group of contrasting previous attainment in physics were conducted, comparing their responses in the post-test with those in the pre-test. In addition, responses from interviews carried out during the teaching were also drawn upon.

All four students developed an understanding of the four aspects of energy targeted in the research-informed teaching sequence. The students with higher previous attainment in physics did not have a sophisticated understanding of the energy concept prior to teaching, though they soon showed themselves capable of using the ideas of store, transfer, conservation and degradation to produce energy descriptions of phenomena both qualitatively and quantitatively. The students with a more modest record of previous attainment in physics also showed themselves capable of producing energy descriptions in terms of store, transfer, conservation and degradation though they both had difficulties in using ideas about conservation and degradation for some time.

DISCUSSION

We set out to design a teaching intervention with the following characteristics:

- It should meet the requirements of the Cyprus National Curriculum;
- It should be internally coherent and consistent with contemporary perspectives on energy;
- Teachers should be able to implement it without the requirement for significant additional training; and
- It should result in significantly improved conceptual understanding amongst students after teaching, compared to the understanding that they might expect to have by following a more typical teaching approach.

At the beginning of this chapter, we described the design of the teaching intervention for energy as research informed. Millar, Leach, Osborne and Ratcliffe (2006) introduced the term *research evidence-informed practice* to describe educational practices where the design is explicitly informed by theoretical and/or empirical evidence from research. In Millar et al.'s terms, research evidence-based practice is practice which might be claimed to be better than other practices in some sense based on evaluation evidence. The first point we wish to make relates to the design decision to adopt a more conceptual approach to energy for students following the Cyprus physics curriculum at this stage in their physics education. This is a decision about an educational intention and, as previously indicated, other designers have chosen to adopt a more materialist approach to energy to address a different educational intention. This decision is at a relatively large grain size, in that it is fundamental to the whole teaching sequence. Our intention was to find out the extent to which it is feasible to introduce to students an account of energy more consistent with that used in contemporary physics, so they can provide coherent accounts of phenomena as required by the official curriculum, and also to illuminate issues faced in teaching and learning using such an approach.

Whatever the outcomes of the evaluation of the implementation, however, we do not believe it is possible to use these as evidence that a more conceptual approach is either better or worse than a more materialist approach. Rather, by being explicit about why this design decision was taken and what was found in evaluating the design, we think it is possible to show future designers some of the opportunities and pitfalls should they decide to use a conceptual approach to energy. As such, we do not aspire to use evidence from this study to suggest the approach is somehow better than others and, therefore, should be adopted (termed 'research evidence-based practice' by Millar et al., 2006). Rather, we are proposing the educational intervention described in this chapter is an example of research evidence-informed practice that will hopefully yield insights useful to future designers of teaching interventions about energy.

The second point we wish to make is, in designing teaching interventions in physics (and probably other subjects), the majority of design decisions are at a fine grain size. We exemplified how the choice of phenomena and events, the appearance of computer simulations, the visual representation of ontological entities such as energy store and transfer pathway, and the communicative approach adopted for teacher-to-whole-class talk were informed by the design tools learning demand and communicative approach for one part of one lesson. Very many such decisions are made by designers working on a teaching sequence. It is impossible to communicate all of these in the space of a single chapter, and yet, the rationale for such decisions is probably the most useful product of design research for future designers. Earlier in this chapter we reviewed research evidence evaluating whether it is productive to think about physics learners' ideas about energy as developing along particular trajectories. Learners' understanding of energy is shaped fundamentally by the accounts of energy that they are exposed to through teaching. In our case, we presented a conceptual approach to energy through the teaching and evaluated students' comprehension. We have presented evidence to show the teaching was, to a large extent, successful in addressing its aims. We have also reviewed literature where a material approach to energy was used in teaching. There is evidence that some of these approaches were also successful in meeting their aims. Rather than focusing on single learning progressions for fundamental concepts such as energy, we think it is more productive to consider students' responses when different design decisions are taken based on fundamental questions about the goals of teaching.

In the previous section of this chapter, we presented evidence to show the research-informed teaching sequence was relatively straightforward for the teacher in this study to implement. As this study is based on a single teacher's implementation of the research-informed teaching sequence means it would be unwise to make general claims about ease of implementation. Nonetheless, the evidence presented earlier in this chapter from both teacher and students does not bring forward any contextual information suggesting this teacher or her students were in any way atypical of others in Cyprus. In line with realist evaluation methods described by Pawson and Tilley (1997), our intention is to produce evidence about what can be achieved when a teacher sets out to address explicit educational goals using a designed intervention in a naturalistic setting. We acknowledge it is likely that more effort has been expended by the teacher in conducting this teaching intervention than would normally be the case. As our intention was to assess the feasibility of enabling school students to use a more conceptual approach to energy, we think this is a legitimate approach. We contend, although underdetermined by the findings of this study, it is plausible that the research-informed teaching sequence could be implemented relatively straightforwardly by other Cypriot physics teachers of similar experience and training.

We also presented evidence in the previous section that the students who had studied the research informed teaching sequence performed significantly better on post-test questions. We further presented evidence in the previous section that the students who had studied the research informed teaching sequence performed significantly better on post-test questions about energy than their peers who had followed a more typical approach to teaching energy. This finding supports the claim that the approach to energy used was coherent, so, it is indeed possible to introduce a conceptual treatment of energy to students of varying prior attainment in physics in such a way that they can make sense of it. The research informed teaching sequence was successful in meeting the aim of promoting significantly improved conceptual understanding among students who studied it, than the understanding they would probably have developed as a result of a more typical teaching approach. The purpose of our study was not to show whether a conceptual or materialist treatment of energy is more accessible to learners. Rather, our claim is that it is possible to introduce students to a conceptual account of energy in such a way that they understand it, thereby promoting an understanding more consistent with that used in contemporary physics. Some studies in the literature have promoted a materialist approach on the grounds of accessibility to students (e.g. Millar, 2005); our findings suggest such a design decision is not inevitable. We see this as an important insight for future designers working on teaching energy with students at this age and stage of their education in physics.

It is not possible for us to attribute particular outcomes in terms of the success of student learning, to particular design decisions embodied in the research informed teaching sequence. We contend, however, that the relative success of students who had followed the research informed teaching sequence compared to their peers following a more typical teaching approach provides supporting evidence that the treatment and sequencing of content in the teaching sequence was coherent to the students. This contention may be of interest to future designers working on teaching energy with students at this age and stage of their education in physics.

APPENDIX A: PRE-TEST QUESTIONNAIRES

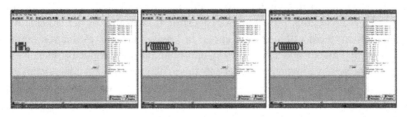

Study carefully the simulation. Afterwards, answer the following questions.

1.   Describe carefully and in detail what you see happening in the simulation-no explanation needed!

_____

_____

_____

2.(a) For each of the events in the simulation described above: explain as best as you can why they occurred.

_____

_____

_____

(b) At the very BEGINNING of the simulation event, is there any energy?
YES or NO? Choose one.
If YES, where?

_____

_____

_____

If NO, why not?

_____

_____

_____

3.   Think now about the END of the simulation event, when the ball has stopped moving. Is there any energy?
YES or NO? Choose one.
If YES, explain.

_____

_____

_____

If NO, why not?

_____

_____

4. What can you say about the amount of energy at the BEGINNING com-
    pared with the amount of energy at the END?

    _____

    _____

5. Finally, describe overall what has happened to the energy of the event from
    BEGINNING to END.

    _____

    _____

6. Look at the following. Which of them has energy? Give a brief explanation.

    _____

    _____

    _____

    _____

    _____

    _____

    _____

    _____

OBJECT	YES/NO	EXPLANATION
1. A moving car		
2. A battery		
3. A runner		
4. A book on a shelf		
5. A barrel of petrol		
6. A stretched elastic band		

APPENDIX B: POST-TEST QUESTIONNAIRES

Part A: Study carefully the simulation. Afterwards, answer the following questions.

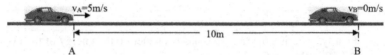

1. Describe carefully and in detail what you see happening in the simulation-no explanation needed!

   _____

   _____

   _____

2.(a) For each of the events in the simulation described above: explain as best as you can why they occurred.

   _____

   _____

   _____

   (b) At the very BEGINNING of the simulation event, is there any energy?
   YES or NO? Choose one.
   If YES, where?

   _____

   _____

   _____

   If NO, why not?

   _____

   _____

   _____

3. Think now about the END of the simulation event, when the ball has stopped moving.
   Is there any energy?
   YES or NO? Choose one.
   If YES, explain.

   _____

   _____

   _____

   If NO, why not?

   _____

   _____

   _____

4. What can you say about the amount of energy at the BEGINNING compared with the amount of energy at the END?

_____

_____

_____

5. Finally, describe overall what has happened to the energy of the event from BEGINNING to END.

_____

_____

_____

PART B

1. A truck is travelling along the road with a constant speed of 5m/s.

At point A, it runs out of petrol and its engine stops working. The car moves for another 10m before finally stopping at point B.

(a) Was any energy stored in the system at point B?

Answer YES/NO

If YES, name that energy and calculate how much there is.

_____

_____

_____

If NO explain why there is no energy.

_____

_____

_____

(b) What has happened to that **amount** of energy at point B? Compare it to that at point A. In what they are similar and in what they differ? Justify your answer.

_____

_____

(c) Describe in words or in any other way and in much detail as you can, what has happened to the energy of the system from point A to point B.

_____

_____

_____

(d) Calculate the force (F) between the wheels of the car and the road in order the car to stop at point B.

_____

_____

_____

2. George makes a bet with his friends Helena and John that he can throw a coin up to a height of 5m.

(a) As the coin moves up through the air it has kinetic energy. Where does it get this energy from?

_____

_____

_____

(b) At the top of its flight the coin is stationary for a moment. Where is the energy now?

_____

_____

_____

(c) Calculate the energy of the coin when it reaches the height of 5m (the mass of the coin is 0.003Kg and g = 10m/s²).

(d) Taking account of the energy at the 5m height, which you have just calculated: calculate the initial speed of the coin as it leaves George's hand.

(e) George throws the coin upwards giving it the initial speed needed to reach the height of 5m and John measures the height. Surprisingly, they see that the coin reached only a height of 4.90m and not at 5m. George repeats the throw of the coin and John measures the height it reaches and its again 4.90m.

Explain why that happens in terms of energy. Use words or any other way in your explanation.

APPENDIX C: FIRST INTERVIEW PROTOCOL

Two classmates, Michael and Niki, leave a crystalline marble to roll on the metallic rail illustrated in the image.

Michael claims that the marble will roll from point A down to point B, hence from point B to point C and then, from point C up to point E.

However, Niki disagrees with Michael and claims that, the marble will roll from point A down to point B, hence from point B to point C and then, from point C up to point D.

1.   With which of the two students do you agree?

2.   Why do you agree with Michael/Niki?

3.   Look at these two diagrams.

   (a) What the rectangles represent?

   (b) What the arrows represent?

   (c) What are the energy stores?

   (d) Do the energy stores exist?

(e) What are the transfer pathways of energy?

(f) Do the transfer pathways exist?

(g) What is the purpose of using the ideas of the energy stores and transfer pathways in physics?

(h) Which diagram do you think describes best the process of the system of the image?

(i) Why do you think is this?

(j) Can you complete the names of the energy stores and the transfer pathways on the diagram?

(k) What can you tell me about the amount of energy at the beginning and the end of the process?

4.   If the marble is allowed to continue its movement, describe how this will be.

5.   Why the marble will do this kind of movement?

6.   Is there energy at the end?

7.   Where energy goes at the end?

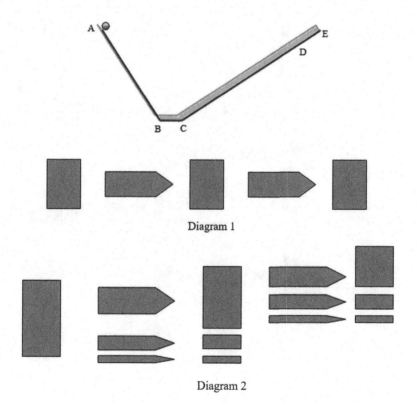

Diagram 1

Diagram 2

APPENDIX D: SECOND INTERVIEW PROTOCOL

The man of the image plays golf on a horizontal field. He hits the ball with his stick and the ball rolls towards the flag and stops beside it.

1.   How do you interpret this physical process?

2.   Look at these two diagrams.

   (a) Which of these do you think that describes best the process of the system of the image?

   (b) Why do you think is this?

   (c) Can you make a full energy description of the process?

   (d) Can you complete the diagram?

   (e) What can you tell me about the amount of energy at the beginning and the end of the process?

   (f) Where the energy goes at the end of the process?

3.   Energy exists?

4.   What is the purpose of using the idea of energy in physics?

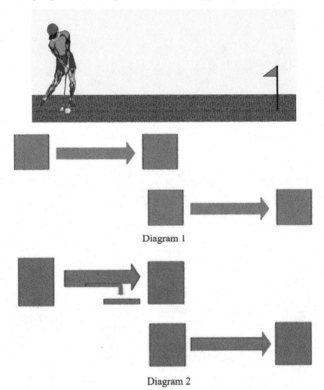

Diagram 1

Diagram 2

REFERENCES

Anderson, G., & Arsenault, N. (1998). *Fundamentals of educational research* (2nd ed.). London, UK: Taylor & Francis.

Ault, C. R., Novak, J. D., & Gowin, D. B. (1988). Constructing vee maps for clinical interviews on energy concepts. *Science Education, 72*(4), 515–545.

Black, P., & Solomon, J. (1983). *Life-world and science-world-pupil's ideas about energy. Entropy in the School. Volume 1.* Roland Eotvos Physical Society, Budapest, Hungary.

Boohan, R., & Ogborn, J. (1996). Differences, energy and change: A simple approach through pictures. *School Science Review, 78*(283), 13–19.

Boyes, E., & Stanistreet, M. (1990). Misunderstandings of 'law' and 'conservation.' A study of pupils' meanings for these terms. *School Science Review, 72,* 51–57.

Brewe, E. (2011). Energy as a substance like quantity that flows: Theoretical considerations and pedagogical consequences. *Physical Review Special Topics. Physics Education Research, 7,* 020106-1–020106-14.

Brook, A., & Driver, R. (1984). *Aspects of secondary students' understanding of energy: Full report.* Centre of Studies in Science and Mathematics Education. University of Leeds. LeedsCampbell, D., & Stanley, J. (1963). *Experimental and quasi-experimental designs for research.* Boston, MA: Houghton Mifflin.

Cohen, L., Manion, L., & Morrison, K.(2007). *Research methods in education* (6th ed.). London, New York: Routledge.

Colonnese, D., Heron, P., Michelini, M., Santi, L., & Stefanel, A. (2012). A vertical pathway for teaching and learning the concept of energy. *Review of Science, Mathematics and ICT Education, 6(1),* 21–50.

Cyprus Ministry of Education and Culture. (2014). *The education system of Cyprus.* Retrieved from http://www.moec.gov.cy

Daane, A. R., Vokos, S. & Scherr, R. E. (2014). Goals for teacher learning about energy degradation and usefulness. *Physical Review Special Topics. Physics Education Research, 10,* 020111-1–020111-16.

Driver, R., & Erickson, G. (1983). Theories-in-action: some theoretical and empirical issues in the study of students' conceptual frameworks in science. *Studies in Science Education, 10*(1), 37–60.

Driver, R., & Warrington, L. (1985). Students' use of the principle of energy conservation in problem situations, *Physics Education, 19*(2), 59–66.

Duit, R., & Haeussler, P. (1994). Learning and teaching energy. In P. Fensham, R. Gunstone, & R. White (Eds.), *The content of science: A constructivist approach to its teaching and learning* (pp. 185–200). Abingdon: Routledge.

Duit, R. (1981). Students' notions about energy concept before and after instruction. In W. Jung, H. Pfundt & C. Von Rhoneck (Eds.), *Proceedings of the International Workshop on problems concerning students' representation of physics and chemistry knowledge,* (pp. 268–319). September 14–16, Pedagogische Hochschule, Ludwigsburg.

Duit, R. (1985). In search of an energy concept. In R. Driver & R. Millar (Eds.), *Proceedings of an invited conference: Teaching about energy within the secondary science curriculum* (pp. 67–101). Leeds, CSSME.

Ellse, M. (1988). Transferring not transforming energy. *School Science Review, 69*, 427–437.

Falk, G., & Herrmann, F. (1977). *Thermodynamik-nicht Wärmelehre, sondern Grundlage der Physik. Teil I: Energie und Entropie [Thermodynamics-not the theory of heat but the principles of physics. Part I: Energy and entropy]*. Hannover: Schroedel.

Falk, G., & Herrmann, F. (1978). *Thermodynamik-nicht Wärmelehre, sondern Grundlage der Physik. Teil II: Das Größenpaar Menge und chemisches Potential [Thermodynamics-not the theory of heat but the principles of physics. Part II: The size of pair population and chemical potential]*. Hannover: Schroedel.

Falk, G., & Herrmann, F. (1979). *Ein moderner Physikkurs fürAnfängér und seine Bergündung [A modern physics course for beginners and its rationale]*. Hannover: Schroedel.

Falk, G., & Herrmann, F. (1981). *Reaktionen in Physik, Chemie und Biologie [Reactions in physics, chemistry and biology]*. Hannover: Schroedel.

Falk, G., Herrmann, F. & Schmid, B. (1983). Energy forms or energy carriers? *American Journal of Physics, 51*(12), 1074–1077.

Feynman, R. P., Leighton, R. B., & Sands, M. (1963). *The Feynman lectures on physics Volume 1*. London, UK: Addison-Wesley, Reading-Mass.

Field, A. (2009) *Discovering statistics using SPSS* (3rd ed.). London, UK: Sage.

Gayford, C. G. (1986). Some aspects of the problems of teaching about energy in school biology. *European Journal of Science Education, 8*(4), 443–50.

Gilbert, J., & Pope, M. (1982). *School children discussing energy*. Report of the Institute of Educational Development, University of Surrey, Guildford.

Gilbert, J., & Pope, M. (1986). Small group discussions about conception in science: A case study. *Research in Science and Technological Education, 4*, 61–76.

Gupta, A., Hammer, D., & Redish, E. F. (2010). The case of dynamic models of learners' ontologies in physics. *Journal of Learning Sciences, 19*, 285–320.

Hobson, A. (2004). Energy flow diagrams for teaching physics concepts. *The Physics Teacher, 42*, 113–117.

Institute of Physics. (2006). *Supporting physics teachers 11–14* (SPT). London, UK: Institute of Physics.

Lawrence, I. (2007). Teaching energy: thoughts from the SPT11-14 project. *Physics Education, 42*(4), 402–409.

Leach, J., & Scott, P. (2002). Designing and evaluating science teaching sequences: An approach drawing upon the concept of learning demand and a social constructivistic perspective on learning. *Studies in Science Education, 38*, 115–142.

Leach, J., & Scott, P. (2003). Learning science in the classroom: drawing on individual and social perspectives. *Science and Education, 12*(1), 91–113.

Mac, S. Y., & Young, K. (1987). Misconceptions in the teaching of heat. *School Science Review, 68*, 464–470.

Millar, R. (Ed.). (2005). *Teaching about energy*. Retrieved from http://www.york.ac.uk/depts./educ/research/.../Paper11Teachingaboutenergy.pdf.

Millar, R., Leach, J., Osborne, J., & Ratcliffe, M. (2006). *Improving subject teaching: Lessons from research in science education*. London: Routledge Falmer.

Mortimer, E., & Scott, P. (2003). *Meaning making in secondary science classrooms*. Buckingham: Open University Press.

Ogborn, J. (1986). Energy and fuel: the meaning of the 'go of things.' *School Science Review, 68*, 30–35.

Ogborn, J. (1990). Energy, difference and danger. *School Science Review, 72*, 81–85.

Orfanidou, D. (2014). *Developing and evaluating research-informed instruction about energy in Cyprus high-schools.* Unpublished Doctoral Thesis, Sheffield Hallam University.

Pawson, R., & Tilley, N. (1997). *Realistic evaluation.* London, UK: Sage.

Ruthven, K., Laborde, C., Leach, J., & Tiberghien, A. (2009). Design tools in didactical research: instrumenting the epistemological and cognitive aspects of the design of teaching sequences. *Educational Researcher, 38*(5), 329–342.

Scherr, R. E., Close, H. G., McKagan, S. B. & Vokos, S. (2012). Representing energy. I. Representing substance ontology for energy. *Physical Review Special Topics. Physics Education Research, 8*, 020114-1–020114-11.

Schmitt, G.B. (1982). Energy and its carriers. *Physics Education, 17*, 212–218.

Scott, P., Leach, J., Hind, A., & Lewis, J. (2006). Designing research evidence-informed teaching approaches. In R. Millar, J. Leach, J. Osborne, & M. Ratchliffe (Eds.), *Improving subject teaching: Lessons from research in science education.* (pp. 60–78). London, UK: Routledge.

Slotta, J. D., & Chi, M. T. H. (2006). Helping students understand challenging topics in science through ontology training. *Cognition and Instruction, 24*(2), 261–289.

Solomon, J. (1982). How children learn about energy-or does the first law come first? *School Science Review, 63*(224), 415–422.

Solomon, J. (1983). Messy, contradictory and obstinately persistent: A study on children's out-of school ideas about energy. *School Science Review, 65*, 225–229.

Stead, B. (1980). *Energy.* LISP Working Paper 17 Science Education Research Unit, University of Waikato, Hamilton, New Zealand.

Stylianidou, F. (1997). Children's learning about energy and processes of change. *School Science Review, 79*(286), 91–97.

Summers, M. K. (1983). Teaching heat-an analysis of misconceptions. *School Science Review, 64*, 670–676.

Van Heuvelen, A., & Zou, X. (2001). Multiple representations of work-energy processes. *Journal of Science Education, 69*(2), 184–194.

Warren, J. W. (1982). The nature of energy. *European Journal of Science Education, 4*(3), 295–297.

Warren, J. W. (1983). Energy and its carriers: a critical analysis. *Physics Education, 18*, 209–212.

Warren, J. W. (1986). At what stage should energy be taught? *Physics Education, 21*, 154–156.

Warren, J. W. (1991). The teaching of energy. *Physics Education, 26*, 8–9.

Watts, D. M., & Gilbert, J. K. (1983). Enigmas in school science: students' conceptions for scientifically associated words. *Research in Science and Technological Education, 1*, 161–171.

Watts, D. M., & Gilbert, J. K. (1985). *Appraising the understanding of science concepts: energy.* Department of Educational Studies, University of Surrey, Guildford.

Wiersma, W. (1986). *Research Methods in Education: An Introduction* (4th ed.). Boston, MA & London, UK: Allyn and Bacon.

Wilkinson, D., & Birmingham, P. (2003). *Using research instruments: a guide for researchers.* Thousand Oaks, California; London: Routledge /Falmer.

Yao, J., Guo Y., & Neumann, K. (2017). Refining a learning progression of energy. *International Journal of Science Education, 39*(17), 2361–2381.

BIOGRAPHIES

Mohan Aggarwal is professor and chairman of physics department at Alabama A&M University. He earned his Ph.D. degree in physics from Calcutta University in 1974 and did his post-doctoral work at Pennsylvania State University in solid state devices. After that he joined as a Research Associate in NASA Spacelab-3 space flight experiment on the growth of infrared crystals in microgravity at Alabama A&M University. He has extensive experience in the bulk crystal growth and characterization of a variety of organic and inorganic nonlinear optical crystals such as divinyl anisole, Schiff base compounds, bismuth silicon oxide, barium titanate, piezoelectric materials such as PMN-PT and scintillator materials using melt growth techniques. He is the author or coauthor of more than 219 publications including two books and five book chapters. He is a member of many learned societies such as American Vacuum Society, American Association of Physics Teachers and SPIE.

David Brown holds an Ed.D. in science education from the University of Massachusetts. He currently is an Associate Professor in the Department of Curriculum and Instruction at the University of Illinois at Urbana-Champaign. His research focuses on the dynamics of instructional interactions in science. This research focus is informed by a complex dynamic systems perspective on the various dynamics involved with instructional interactions, including social, affective, and

Physics Teaching and Learning: Challenging the Paradigm, pages 247–254.
Copyright © 2019 by Information Age Publishing
All rights of reproduction in any form reserved.

particularly sense making dynamics. Instructional contexts include classroom instruction, tutoring, and technology assisted instruction.

Isaac Buabeng holds a Ph.D. in science education from University of Canterbury, New Zealand. He is an experienced Senior Lecturer and researcher in science education at the Department of Basic Education, University of Cape Coast, Ghana. Prior to becoming a teacher educator, Isaac was a Junior and High School science, mathematics and physics teacher. Isaac's research has focused on sex differences in science achievement, females' participation in physics studies at tertiary levels and innovative approaches to science teaching, with physics in particular, at primary, secondary and tertiary levels. As part of his doctoral thesis, Dr. Buabeng investigated whether tertiary study adequately prepared and allowed pre-service teachers to become effective in in classroom teaching. He has research experience in both Ghana and New Zealand and has authored and published several articles in refereed journals.

Ted Clark is an Associated Professor in the Department of Chemistry and Biochemistry at The Ohio State University (OSU). Dr. Clark earned his Ph.D. in Chemistry from the University of Michigan in Ann Arbor. Following a post-doctoral research position with Dr. Philip Grandinetti in the area of solid-state nuclear magnetic resonance, Dr. Clark has focused on teaching and learning in undergraduate courses at OSU. He has been involved in Modeling Instruction professional development workshops for more than a decade with an emphasis on high school physical science instruction. His research interests include implementation and evaluation of active learning strategies in large-enrollment STEM courses and inclusion of authentic research experiences in laboratory courses.

John Clement holds a doctorate in science education from the University of Massachusetts, Amherst and currently is Professor Emeritus of Science Education there. His research interests are in thinking in scientists, imagistic thinking, and fostering conceptual change in classrooms. Dr. Clement has authored four books, and over 70 chapters and articles. He was the recipient of the Distinguished Contributions to Science Education Through Research Award from the National Association for Research in Science Teaching. He has served on boards for the National Science foundation, and the National Science Board.

Lindsey Conner is the Dean in the College of Education, Psychology and Social Work at Flinders University, Australia. Prior to this she was Deputy Pro Vice Chancellor at the University of Canterbury, NZ. Her research has focused on students' learning in science and technology and teacher education. Prof Conner led the 7 country Pacific Circle Consortium project on Teacher Education for the Future. In 2013 she was a consultant to NIER (Japan) working with the Ministry of Education on infusing competencies across curriculum, and a fellow in residence at Southeast Asian Ministers of Education Organization Regional education centre for science and mathematics, Penang Malaysia. Prof Conner has developed courses for mentoring teacher educators from universities in Bangladesh, China, Malaysia and Korea. Previously, Prof Conner was the New Zealand coordinator for the OECD Innovative Learning Environments Project and Commissioner for the New Zealand Olympic Education Committee.

Lin Ding, Ph.D. is an Associate Professor of STEM Education in the Department of Teaching and Learning at The Ohio State University. Dr. Ding's scholarly interests lie in discipline-based science education research. His work includes theoretical and empirical investigations of learners' content learning, problem solving, reasoning skills, and epistemological development. Dr. Ding has been leading or co-leading several federal and state projects sponsored by the National Science Foundation and the Ohio Department of Education. Also, he has served as Associate Editor for the PER-Central resource center and Editor for the Physics Education Research Conference Proceedings.

Haim Edri is a Ph.D. candidate at the Department of Science Teaching at the Weizmann Institute of Science. He is an experienced physics teacher focusing on project-based learning of computational physics adapted to the level of high school students. He designed and taught the 10th grade course of the Interdisciplinary Computational Physical Science program.

Lisa Hardy completed her Ph.D. in Education at UC Davis, and is currently a researcher at the Concord Consortium in Emeryville, CA. Dr. Hardy's research interests are in designing and developing technology-enhanced learning environments for K–12 and undergraduate STEM classrooms. Her current research investigates how "Maker" technologies such as Raspberry Pis and "Internet of Things" sensors can engage students in authentic science inquiry. Her dissertation research explored how university physics students engage in scientific reasoning when interacting with one another around networked simulations. She has a double B.S. in Physics and Biochemistry/Molecular Biology from the University of California, Davis and an M.S. in Physics from the University of Wisconsin.

Kathleen A. Harper is a Senior Lecturer in the Department of Engineering Education at The Ohio State University. She earned her Ph.D. in physics, specializing in physics education research, from Ohio State under the guidance of Alan Van Heuvelen. She has been involved in the national Modeling Instruction movement for over 20 years and has directed Modeling workshops in central Ohio for 14 years. She frequently presents workshops and seminars of a variety of topics in physics and engineering education research, with a particular focus on the link between educational research and practical classroom tools and techniques. She is active with both the American Association of Physics Teachers and the American Society for Engineering Education.

J.W. Harrell is Professor Emeritus of Physics at The University of Alabama. He received his PhD in physics from the University of North Carolina. He is a former department chair and an Arts & Sciences Teaching Fellow. Dr. Harrell's research interests include magnetic materials and physics education, and he has published more than 100 refereed papers on these topics. As a past chair of the IEEE Magnetics Society Education Committee, he was instrumental in the development of annual international summer schools for graduate students in magnetics. He was a leader in the implementation of Studio Physics at UA. He has been active in several programs to graduate new high school physics, chemistry, and mathematics teachers and to provide in-service training for practicing high school physics and physical science teachers. Recent and on-going teacher education programs in which he has participated include PhysTec, PS-21, NSF-Noyce, and NSF-APEX.

Elon Langbeheim is a research scientist at the Department of Science Teaching at the Weizmann Institute of Science. He received his Ph.D. in Science Teaching from the Weizmann Institute of Science in 2014. Then, he held postdoctoral fellowships at the University of Haifa in 2013–2014 and at the Mary Lou Fulton Teachers College at Arizona State University in 2014–2016. His research focuses on introducing complex physical science topics to secondary school students as well as on the professional development of science teachers. He developed and taught the 12th grade course in the Interdisciplinary Computational Physical Science program described in the chapter.

John Leach completed his PhD in science education at The University of Leeds (UK), where he later went on to become Professor of Science Education. His research focuses on teaching and learning science in formal settings. He has held research grants from funders in the UK and Europe, and has published extensively in international journals focusing on science education. He has served on many editorial boards and is a longstanding member of the editorial board of the journal *Research In Science Education*. He was a member of the 2008 Research Assessment Exercise sub-panel, and deputy-chair of the 2014 Research Excellence Framework sub-panel for education in the UK, and is deputy-convenor of the 2020 Research Assessment Exercise in Hong Kong. Professor Leach has served as a Dean and Pro Vice-Chancellor at three English universities before taking up his current role as Director of the Institute for Childhood and Education at Leeds Trinity University.

Justina Ogodo, Ph.D. in science education from The University of Alabama, is a post-doctoral researcher in STEM Education in the Department of Teaching and Learning at The Ohio State University. Her research focuses on science curriculum and instruction, STEM teacher pedagogical content knowledge and urban education/culturally responsive teaching. She uses her experience in STEM education to provide pre-service teachers with effective tools to prepare them for the profession. She mentors early career teachers through induction programs and provides targeted professional development to enhance in-service teachers' content knowledge and pedagogical skills. In addition to presenting peer-reviewed papers at annual meetings of several professional associations, AERA, NARST, & AAPT; she has authored two book chapters: and some articles.

Dora Orfanidou holds a BSc in Physics from the University of Patras, Greece, an MA in Learning in Natural Sciences from the University of Cyprus, a PhD in Science Education from Sheffield Hallam University, UK and a Physics Teacher Certificate from the Cyprus Pedagogical Institute. She is currently a senior physics teacher in a public high school in Cyprus. Her research interests are in two related areas. The first area is the development of innovative teaching approaches to introduce physics concepts at the secondary school level, including through the medium of ICT. The second area is the development of physics curricula at the secondary level to establish and promote effectively the development of scientific concepts. Dr Orfanidou presented research work at various Cyprus and key international conferences such as GIREP and ESERA.

Sam Safran received his Ph.D. in Physics from MIT, followed by a postdoc at Bell Laboratories. From 1980–1990 he served as a Senior Staff member in the Complex Fluids Physics group of Exxon Research and Engineering in New Jersey. Prof. Safran joined the faculty of the Weizmann Institute of Science in 1990 and is currently in the Department of Chemical and Biological Physics. His current research interests in the theory of soft and biological matter focus on the physics of mechanobiological structure, dynamics and function and in science education in these areas. Honors include the de Gennes Award of the European Physical Journal and the Beller Lectureship of the American Physical Society. He is the author of a graduate text on the physics of surfaces, interfaces and membranes, and served as the scientific advisor to the Interdisciplinary Computational Physical Science program.

Nava Schulmann is a research engineer at INRIA, Strasbourg, France. Her work focuses on stochastic filtering, parameter estimation and system observability in the context of real-time simulation for pre-operative guidance. Her Ph.D. (2009–2012), at the theory and simulation group of the Charles Sadron Institute in Strasbourg focused on spontaneous formation of structures and universal scaling behaviors in polymer and fiber systems. As a post-doctoral fellow at the Weizmann Institute of Science (2013–2015), she participated in building the foundations of the Interdisciplinary Computational Physical Science program.

Marsha Simon earned her Ph.D. in educational research from the University of Alabama. She is currently an assistant professor of educational research at the University of West Georgia. Her research interests include the impact of physics professional development, and the imposter syndrome and the role of Black women in STEM fields. An early career scholar, Dr. Simon has refereed articles in the *Personality and Individual Differences, School Science and Mathematics and Middle Grades Research Journal*. She has recently presented papers at the annual meetings of the American Educational Research Association (AERA), the National Association for Research in Science Teaching (NARST) and the American Association of Physics Teachers (AAPT). Her professional activities include reviewing for *Personality and Individual Differences* journal as well reviewing submissions yearly for the annual AERA meeting.

Marilyn Stephens holds a Ph.D. in science education from the University of Alabama. She is a National Board-Certified science teacher and has taught both secondary and college course work for more than 35 years. She currently is doing research in physics education with the University of Alabama. Her research interests are in formative assessment, science reform, and induction programs for novice science teachers. Dr. Stephens has recently presented papers and symposia at annual meetings of several professional organizations including the American Association of Physics Teachers, Southeastern Association of Teacher Education, the National Association for Research in Science Teaching and Alabama Science Teacher Association. Among her other professional activities, she has served as a board member of Alabama Science Teachers Association.

Cynthia Szymanski Sunal completed her Ph.D. at the University of Maryland and is now Professor and Chair of Curriculum and Instruction at The University of Alabama, and Director of the Office of Research on Teaching in the Disciplines. She primarily works with elementary and middle level teachers. Among her publications are numerous books, journal articles, and monographs. In addition, she serves as Executive Editor of two journals and publishes a research series. She has also been involved in several funded projects from the National Science Foundation, the Department of Energy, and other agencies. Recently, Dr Sunal received the University of Alabama's College of Education's *Bryant Professor of Research Award*.

Dennis W. Sunal received a Ph.D. from the University of Michigan. He currently is Professor of Science Education at the University of Alabama, Tuscaloosa. His university teaching experiences include undergraduate and graduate courses in physics, engineering, curriculum and instruction, and science education. He holds both secondary (6th–12th) and elementary (K–6th) teacher certifications, and has taught at both levels. His research interests include effective undergraduate science teaching, physics teacher education, and pedagogical content knowledge of science teachers and faculty. He has been project director and co-director on numerous grants (e.g. NSF, NASA, Department of Education, USIA, and U.S. Department of Energy). In addition, Dr. Sunal has published numerous articles and chapters in refereed journals and books. He is also Co-editor of the *Research in Science Education* (RISE) series published by Information Age Publisher.

Tobin White holds a Ph.D. in Education from Stanford University. He is currently an Associate Professor and Chancellor's Fellow in the School of Education at UC Davis. Dr. White's research focuses on investigating innovative uses of technology in the teaching and learning of mathematics and science. He has a particular interest in using handheld and tablet computers to support student participation and engagement in novel and interactive forms of STEM classroom activity. Using a design-based research approach, he develops collaborative problem-solving tools and activities in order to investigate intersections between conceptual and social dimensions of learning. A former high school mathematics teacher himself, he has also worked for more than a decade in teacher preparation.

Robert Wallon is currently a doctoral candidate in the Department of Curriculum and Instruction at the University of Illinois at Urbana-Champaign. He has served as a research assistant on the GRASP project, and his research interests focus on educational technologies for science teaching and learning. As a former high school science teacher, he values research that considers implications for practice and that acknowledges the complexities of authentic learning environments.

Grant Williams holds a B.Sc. in physics from Mount Allison University, a B.Ed. in science education and a M.Ed. in educational administration from the University of New Brunswick, and an Ed.D. in mathematics, science, and learning technologies from the University of Massachusetts Amherst. He is currently Director and Associate Professor in the School of Education at St. Thomas University in Fredericton, Canada. His research interests are in cognition, model-based science teaching and learning, and classroom discourse. Dr. Williams developed the online resource www.Kinulations.com (Williams, G. (2013). *Kinulations: Minds-on, bodies in science learning*) which provides lesson plans and video exemplars for teachers of K–12 science interested in engaging students in *Kin*esthetic Simu*lations* as a means of understanding abstract scientific concepts. In addition to publications in the International Journal of Science Education and Science Scope, he has recently presented papers at the National Association for Research in Science Teaching, American Educational Research Association, and Canadian Society for Studies in Education annual conferences.

David Winter is a lecturer in the School of Teacher Education at the University of Canterbury, Christchurch, New Zealand. He is coordinator for the Graduate Diploma in Teaching and Learning (Secondary) programme. He teaches courses in professional studies and science education and has research interests in Physics and Chemistry teaching. David has delivered professional development courses for both in-service teachers and teacher educators from developing countries. Prior to becoming a teacher educator, David was for many years a forensic scientist and he enjoys using forensic science as a context for teaching nature of science themes.

Edit Yerushalmi heads the Physics Education Research Group at the Science Teaching Department, Weizmann Institute of Science. She a holds a Ph.D. in science education from the Weizmann Institute of Science and an M.Sc. in physics from the Technion—Israel Institute of Technology. Prof. Yerushalmi serves as the scientific advisor for the national teacher center in Physics. She directs the "Research Projects in High School Physics" teacher preparation program in which physics teachers are trained to lead research projects with their students. In addition, she directs the "Gateway to Physics" project

Printed in the United States
By Bookmasters